Butterflies

A Complete Guide to Their Biology and Behavior

Dick Vane-Wright

Comstock Publishing Associates
a division of
Cornell University Press
Ithaca, New York

For Hazel

First edition © The Trustees of the Natural History Museum, London, 2003
Second edition © The Trustees of the Natural History Museum, London, 2015

The Author has asserted his right to be identified as the Author of this work
under the Copyright, Designs and Patents Act 1988.

Second edition first published in the United States of America in 2015 by
Cornell University Press
First printing, Cornell Paperbacks, 2015
Simultaneously published in the United Kingdom by the Natural History
Museum, London

Librarians: A CIP catalog record for this book is available from the Library
of Congress.
ISBN 978-1-5017-0017-0 (pbk. : alk. paper)

Designed by Mercer Design, London
Reproduction by Saxon Digital Services
Printed in China by C&C Offset Printing Co., Ltd.

Front cover: Red admiral butterfly, *Vanessa atalanta* © Ernie Janes/Photoshot
Back cover: Swallowtail butterfly, *Papilio thoas* © The Trustees of the Natural
History Museum, London

Paperback printing   10 9 8 7 6 5 4 3 2 1

Contents

Introduction

BUTTERFLIES AND MOTHS BELONG to the order Lepidoptera. This is one of the major groups of insects, and members of the Lepidoptera are recognizable by the broad, flattened scales that cover both pairs of wings and, in all but a few of the most ancient sorts, possession of a coiled feeding tube or proboscis. Although not as diverse as flies and midges (order Diptera), beetles (order Coleoptera), or bees, wasps and ants (order Hymenoptera), the Lepidoptera include more than a quarter of a million different species. Like bees, flies and beetles, the butterflies and moths have a life cycle involving a complete change in physical form (metamorphosis). The worm-like caterpillar that hatches from the egg feeds voraciously, using its chewing jaws, until it is fully grown. It then stops feeding and turns into an immobile stage: the pupa or chrysalis. Within the chrysalis the caterpillar body is radically transformed to produce the adult that eventually emerges, with its drinking-straw proboscis and beautiful wings.

OPPOSITE Silver-studded blues, *Plebejus argus*, roost communally at night. Pictured here near Antwerp, Belgium, these three are covered in early morning dew. Whether or not butterflies really sleep is an open question.

BELOW A small part of the hindwing of an Indian peacock swallowtail, *Papilio krishna*. The brilliant green or lilac of each dimpled scale is produced by its special layered structure; the reddish scales contain a melanin pigment.

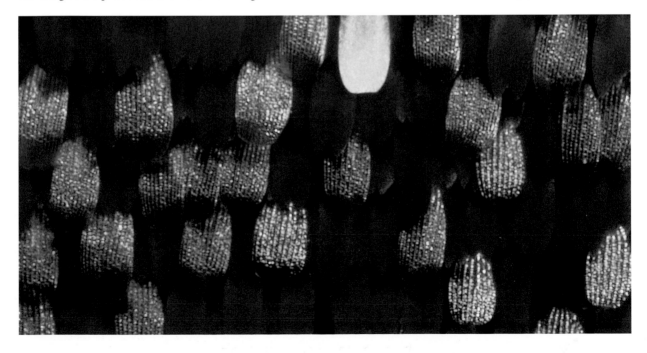

WHAT IS A BUTTERFLY?

The butterflies comprise a conspicuous subgroup of the Lepidoptera. Traditionally they have been divided into two supposedly distinct sorts: true butterflies (superfamily Papilionoidea) and skippers (superfamily Hesperioidea). According to this scheme, the true butterflies have broad, flapping wings, and include familiar insects such as swallowtails, whites, blues, browns, monarchs and admirals. Most skippers have a wider head but narrower, more moth-like wings that beat very fast, giving them a rapid, darting flight. However, the relatively recent discovery that a rather obscure American family of moth-like Lepidoptera, the Hedylidae, belong to the butterflies has led to an ongoing reappraisal of the entire group. Currently the Hedylidae are seen as most closely related to the skippers, with the swallowtails, Papilionidae, as the oldest subgroup of all – so the new proposal is that all butterflies, including the skippers and the Hedylidae, should now be included in a single superfamily, to be known as the Papilionoidea.

BELOW **This Essex skipper,** *Thymelicus lineola,* **is very alert. Typical features of most of the 4,000 or so species of skippers include a broad head, separated wings at rest, and a stout body.**

LEFT A large white, *Pieris brassicae*, at rest. All four of its broad wings are held together upright over its 'back' in typical true butterfly fashion.

LEFT Head-on view of an emperor moth, *Saturnia pavonia*, with its feathery antennae. The antennae of butterflies are always simple, never 'feathered'.

The frequently repeated question, 'What's the difference between a moth and a butterfly?', is rather like asking how to separate novels from books. The butterflies, which probably include about 20,000 species, are in reality just one amongst the many subgroups of moths. At least four of the other superfamilies within the Lepidoptera are more numerous in species, and all are regarded as moths. These are the geometroids ('inchworm' or 'looper' moths, with well over 20,000 known species), the noctuoids ('owlet' moths, with nearly 35,000 species already named), the pyraloids (containing at least 30,000 moth species, many still unnamed) and the gelechioids (a huge group of mostly small moths, many unnamed, which could eventually exceed 40,000 species).

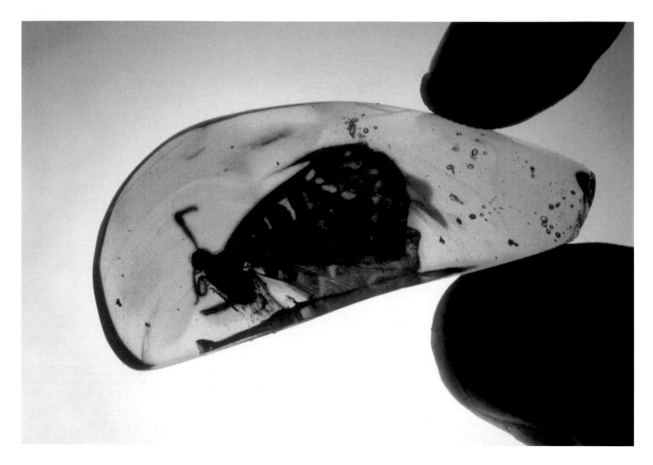

ABOVE **A beautifully preserved metal-mark butterfly (family Riodinidae) in a piece of Dominican amber, about 20 million years old. The small foreleg that is typical of the family is clearly visible, together with the partly uncoiled proboscis.**

A combination of characters separates butterflies from other Lepidoptera. Most butterflies fly by day, though a very small number fly at night, including the Hedylidae, while many more, including the 'evening browns', fly actively only at dusk. Butterflies broad, overlapping wings are not directly hooked together (male regent skippers being the lone exception), and the antennae are the same in both sexes, often being thickened towards the tip to make a 'club' but never 'feathered'. Generally butterflies rest with their wings held together above the body, or held out flat. By contrast, most moths fly at night, have wings that are directly connected together by a hook-and-eye type of mechanism, and their antennae, which are often feathered, are usually larger or thicker in the males. Many moths rest with their wings folded tent-like above the abdomen.

Characters that more reliably identify butterflies, separating them from other Lepidoptera, are found in rather obscure features of the brain, muscles and external skeleton. In recent years analysis of DNA data has proved consistent with the idea that the butterflies, including the Hedylidae, do form a 'natural group' nested within the numerous other groups of Lepidoptera, such as the gelechiids and burnet moths, but not closely related to moths such as inchworms, hawks and owlets.

HOW OLD ARE THE BUTTERFLIES?

Very few butterfly fossils are known, under 100 according to a recent catalogue. The oldest specimen that can be placed as a butterfly with reasonable confidence is an unnamed skipper from a Late Thanetian (Upper Palaeocene) deposit found at Fur Island, Denmark. By that time, about 56 million years ago, the single ancient landmass of Pangaea had already broken up to form the major continents we see today.

The oldest named fossil butterflies are two swallowtail-like species from a Middle Eocene deposit in Colorado: *Praepapilio colorado* and *Praepapilio gracilis*. Estimated to be more recent than the Fur Island skipper, at about 48 million years old, these *Praepapilio* are particularly interesting. The Papilionidae (swallowtails) are now thought to represent the earliest branch of the living butterfly 'family tree'. Some have suggested that *Praepapilio* is related to what is probably the oldest living butterfly lineage, the so-called Archaic swallowtail, *Baronia brevicornis* (see Appendix). However, as pointed out by taxonomist and biogeographer Rienk de Jong, one has to be very careful in making such comparisons and drawing hasty conclusions. The *Praepapilio* fossils do have some of the features that separate the swallowtails from moths and the rest of the butterflies. However, they lack other features that all living Papilionidae have. Thus all we can say, if correctly interpreted on the basis of their preserved wing venation, is that *Praepapilio* represents a lineage older than any living butterfly. So does this mean that the butterflies evolved in the Palaeogene – a geologic period that started 66 million years ago? No. All it does confirm (if correctly interpreted), in conjunction with the Fur skipper, is that the minimum age of our living butterflies is 56 million years.

Amongst the Lepidoptera, butterflies seem to be younger than swift and burnet moths, but older than inchworm, owlet and hawk moths – with the likelihood that the order as a whole is just old enough to have been present on Pangaea before fragmentation proceeded very far. Primitive moths have supposedly been identified from amber deposits of Lower Cretaceous age, dating from about 135 million years ago. There is a view, however, that the major diversification of the Lepidoptera did not occur until the Palaeogene – by which period, as indicated above, it seems pretty certain the butterflies already existed. Once the DNA of many more species has been studied in detail, molecular biology may be able to give us a better idea about the evolution of the whole group, including the age of origin of the butterflies – currently thought to have occurred approximately 100–70 million years ago.

BELOW Named in 1878 from the Late Eocene Florissant Beds of Colorado as *Prodryas persephone*, this extinct butterfly is about 35 million years old. It is considered to belong to the family Nymphalidae. Note the remarkable preservation of the clubbed antennae and wing pattern, as well as the wing veins.

CHAPTER 1
Becoming

THE BEAUTIFUL FLUTTERING INSECTS THAT we know as butterflies are the culmination of a lengthy and complex life cycle. Growth begins with an egg and progresses through several larval stages until the final metamorphosis into a butterfly. Things do not always go smoothly, however, and the insects face many hazards as they strive to reach adulthood.

GETTING STARTED

Find a stinging nettle patch in a sunny spot during May. If you look carefully under the topmost leaves at the edge of the patch, you may find a cluster of 100 or more little ribbed green barrels. These are the freshly laid eggs of a small tortoiseshell, *Aglais urticae*. After about a week, the eggs, each one only about 0.8 mm in diameter, turn yellow and then darken. Tightly curled up within each egg is a very small caterpillar, less than 1.5 mm long. The relatively large head, packed with muscles to power the all-important jaws, becomes clearly visible through the semi-transparent shell. After 7–21 days, depending on conditions, the larva is ready to use those jaws for the first time, to gnaw through the top of its tough shell.

Other than the necessity of gnawing a hole to get out, the small tortoiseshell does not eat its eggshell, and this appears to be the case in many butterflies. However, the females of some species cover their eggs in a layer of 'spumaline', a carbohydrate-rich material, and in such a case the first thing that the little caterpillar will eat is its own eggshell. If a newly emerged larva of the speckled wood, *Pararge aegeria*, is prevented from doing this, it will die. What is it that the young caterpillar must get from the eggshell to survive? Almost certainly some key nutrients, or an inoculation of bacteria or other micro-organisms necessary for normal growth – but precisely what is so essential remains unknown. The females of other species plaster their eggs in what appear to be protective or nutritive scales, and in some cases these are eaten, too.

The eggs of butterflies come in many shapes, with a 20-fold range in volume. Most swallowtails have large, round and smooth eggs, up to 3 mm or more in

OPPOSITE **The gregarious caterpillars of *Morpho telemachus* are as spectacular as the huge 'Morpho' butterflies they will eventually become. These butterflies were photographed at Manu National Park, Peru.**

BELOW **A batch of about 100 eggs laid under a stinging-nettle leaf by a peacock butterfly, *Aglais io*. These eggs are very similar to those of the small tortoiseshell and its North American relative, the fire-rim tortoiseshell, *Aglais milberti*.**

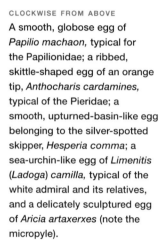

CLOCKWISE FROM ABOVE
A smooth, globose egg of *Papilio machaon,* typical for the Papilionidae; a ribbed, skittle-shaped egg of an orange tip, *Anthocharis cardamines,* typical of the Pieridae; a smooth, upturned-basin-like egg belonging to the silver-spotted skipper, *Hesperia comma*; a sea-urchin-like egg of *Limenitis* (*Ladoga*) *camilla,* typical of the white admiral and its relatives, and a delicately sculptured egg of *Aricia artaxerxes* (note the micropyle).

diameter. The eggs of whites and sulphurs are ribbed and look like little skittles. If you come across a butterfly egg that looks a bit like a miniature sea urchin, it will be a member of the white admiral group. Many of the blues have doughnut-shaped eggs with exquisite sculpturing. Certain skippers have eggs like little pillboxes or pudding-basins, with a finely latticed surface. The significance of the variations in size, and especially in shape and surface structure, is not well understood. However, all have at least one thing in common: at the top, usually in a slight depression, there is a hole or small group of holes. This is called the micropyle and is where sperm pass into the egg during fertilisation, before each one is laid.

EATING IS A SERIOUS BUSINESS

Having escaped from its eggshell, the small tortoiseshell caterpillar has an unremitting diet of nettle leaves ahead of it. The difference in weight between a 'newborn' caterpillar and the fully grown larva is typically 1,000-fold or even more. All that growth must come from what the caterpillar eats. The caterpillar needs to make sure that enough proteins, fats, carbohydrates and vitamins are stored so that when it is ready to pupate, it will be able to mature into a normal-sized, fully functional adult. If it does not eat quite enough, it will turn into a very small adult butterfly that is unlikely to have many offspring. If it eats still less, then it will fail to complete its development altogether.

FAR LEFT The caterpillars of many butterflies are solitary, but those of certain species feed in groups, especially when they are young. These first instar small tortoiseshells have spun a dense silken tent within which they feed communally.

LEFT This first instar Bath white caterpillar, *Pontia daplidice*, has just hatched and is busy devouring its own eggshell. The black, glossy head is very conspicuous.

So eating is a very serious business for a caterpillar. In many species, so long as the food supply is good and continually available, a caterpillar will eat almost non-stop. In this way the larva of a tropical swallowtail, in ideal conditions, can complete its growth in two weeks, more than doubling its weight every two days. If that does not sound impressive, think what would happen to an average newborn human baby if it could grow that fast: it would weigh over a tonne within a fortnight.

STARTS AND STOPS ALONG THE WAY

In almost all parts of the world there are times of year that are not good for growth: it may be too cold, too hot or too dry, or there may be no food. Depending on the species, any stage in the butterfly life cycle can 'sit out' the bad times by some form of hibernation or aestivation. The life processes are almost closed down and it doesn't move, feed or grow. In cooler areas of the world, many species pass the winter as caterpillars in such a quiescent state, waiting for the spring. Once they have entered hibernation, the insects often need a reliable environmental cue, such as increasing day length, to trigger a return to active life.

In dry regions many butterflies get through drought in what is known as 'pupal diapause'. In the chrysalid stage it seems relatively easy to vary how long individuals can stay in suspended animation. If several chrysalids in pupal diapause are placed in cooler and relatively humid conditions, some will usually give rise to butterflies within weeks, but others may stay as pupae for months, or even one or two years. Fifteen years is the record! How such diapause is ended is uncertain, but the significance of the variation almost certainly relates to survival in the unpredictable climate of dry regions, such as deserts.

Caterpillars can also have a daily rhythm of feeding activity. Some relatively slow-growing species feed at night, only becoming active after dark. Before dawn they return to a hiding place to pass the day, which may be too hot, too dry or simply too dangerous for active feeding. Other species feed only by day, and yet others change their daily feeding routine as they get older.

ABOVE This final instar blue-banded king crow, *Euploea eunice*, will soon be ready to pupate. The function of the remarkable curly filaments is a mystery.

STAGES OF GROWTH

One type of periodic interruption to feeding affects all butterflies. As a caterpillar develops, its skin (more correctly the external skeleton, or exoskeleton) does not grow with it, but can only stretch to accommodate the steadily enlarging body within. There is a limit to how much the skin can stretch. As its limit is reached, moulting hormones (ecdysones) are released within the caterpillar to prepare it for a crucial event. A new exoskeleton starts to form under the old one. The caterpillar stops feeding and the old skin splits open. In its new but very soft exoskeleton, the caterpillar struggles out, and the almost dry husk of the old skin falls away. The new skin is then inflated before it hardens and has the potential to stretch to a greater, but still limited, extent. Feeding recommences and continues until the new stretch-limit is reached, and then the whole skin-shedding process is repeated. The caterpillars of most butterflies go through this moulting procedure (ecdysis) four times. The caterpillar between each step is referred to as a 'stage', or instar.

Most butterflies thus have five larval instars, but some have only four, and a few have as many as nine. Regardless of how many instars there are, the stage at which the growth of the caterpillar comes to an end is called the final instar. When all is ready, this final stage caterpillar goes through one last larval ecdysis – but this is very different to those that were necessary to accommodate growth.

THE REMARKABLE TRANSFORMATION OF METAMORPHOSIS

Up until the last larval stage it is only necessary to moult to allow growth to continue. To do this not only is a moulting hormone released, but the caterpillar is also perfused with a juvenile hormone. On entering the final instar, however, the amount of juvenile hormone produced is greatly reduced. This acts as a signal so that under the last larval skin, instead of an even bigger caterpillar, the exoskeleton of the chrysalis forms. When the old larval skin is cast off, the doll-like pupa is revealed. A close look will show where the eyes of the butterfly are forming, along with its antennae, proboscis, legs, wings and abdomen – a sort of shrink-wrapped adult.

Some people have viewed insect metamorphosis, from creeping caterpillar to soaring butterfly, as mysterious or even miraculous. How does it happen? Although very complex in detail, the basic process is simple. During the development of any animal, the single original cell, the fertilized egg, divides, and then its two daughter cells divide again to make four, and so on. Repeating such a process just 24 times

would produce over 16,000,000 cells. To make a mammal such as ourselves takes about 30 successive divisions. As the divisions take place, each cell becomes more and more specialized for its particular function, dependent on exactly where it fits into the growing body: brain cell, heart muscle cell, liver cell, hair follicle cell, and so on. A very similar process of cell differentiation occurs inside the fertilized egg of a butterfly, to produce the little first-instar larva, with all its different functional parts. There is, however, one important difference.

Very early on in the life of the caterpillar, some cells that have made only a few divisions are set aside in small groups, and undergo little or no further development while the caterpillar completes its growth. These cells are destined to develop into various parts of the adult butterfly. After the final instar, when the larva has shed its skin for the last time, many of the caterpillar's cells, such as those

ABOVE Transformation from caterpillar to chrysalis in *Aglais io.* Hanging by its tail, the spiny caterpillar skin splits and the immobile green pupa wriggles free. It remains attached by tiny hooks fixed into a pad of silk spun by the larva.

BELOW The chrysalids of some butterflies, such as this spectacular dwarf crow, *Euploea tulliolus,* have golden reflections. Early butterfly fanciers were called 'aurelians' as a result.

forming its muscles, die and are broken down to form a nutritious 'soup' that fills the central space of the pupa. If you cut open a pupa that is a few days old, this is what you will see ooze out. Meanwhile, the little groups of largely undifferentiated cells, called 'imaginal buds', now start to divide. They use the caterpillar-cell-soup in the same way that a hen's developing embryo uses the egg yolk: as a source of materials and energy. So the metamorphosis is not a direct transformation of the caterpillar – it is more like a new construction using recycled materials. However, many of the larval structures, such as the gut, do pass from the caterpillar to the adult, although they are highly modified in the process.

EMERGENCE OF THE ADULT BUTTERFLY

A The chrysalis of this sulphur butterfly hangs from a silken 'girdle' passing over the developing wings, now clearly visible within.

B The pupal case has split, and the butterfly starts to struggle out. Three of its pinkish-red legs are already free.

C Free at last! But flight is still not possible: the wings are still limp and useless bags.

D A few more minutes of wing expanding and drying, and this clouded yellow, *Colias croceus*, will be ready to go.

When the butterfly is ready to emerge, the pupal case splits open. First the head, legs and antennae appear, then the wings, and finally the abdomen is pulled out from within. At first the little butterfly is rather a sorry- looking affair, crumpled and damp, the soon-to-be beautiful wings hanging like limp bags at each side. The creature clings to a stem or whatever means of support it has, and creeps upwards to find a suitable spot to complete its transformation. At this time the butterfly may release a squirt of meconium, an orange or red slurry of waste products that were stored throughout pupal development (as the pupa is completely enclosed, with no vent to the outside).

In a matter of minutes or hours, depending on circumstances and species, this final stage of the seemingly miraculous process is complete. The wings have expanded, driven by pressure of blood-like fluid (haemolymph) within. Once the wings have dried and hardened, without any hesitation other than some initial fluttering, the butterfly suddenly flaps its wings and sets off on its maiden flight, fully pre-programmed in the art of aerial navigation. After mating, the metamorphic life cycle can begin anew.

A

B

C

D

HOW LONG DOES A BUTTERFLY LIVE?

LEFT Danger is always close. This brilliant male bird wing, *Ornithoptera priamus*, has blundered into the web of a huge *Nephila* spider. Its life is almost over, to be sucked dry by the dark destroyer.

It is often said that a butterfly lives for only a day. In reality, unless overcome by an accident or killed by a predator, most adult butterflies live for several days, commonly two or three weeks, although some live for two to three months or even longer. Those temperate species that pass the winter as adults, such as the mourning cloak, *Nymphalis antiopa*, can live even longer, as much as six months or more, but most of this long life is inactive. In Sulawesi during July 1985, I put a unique mark on a little brown butterfly, *Orsotriaena jopas*, and encountered it 46 days later, looking as fresh as when I first saw it. However, after three weeks or more, most butterflies begin to show their age: their colours start to fade and their wings get rubbed and frayed from encounters with other butterflies, vegetation, spiders' webs and all manner of little accidents. In the adult butterfly, other than those processes involved with reproduction, there is little or no cell division and no growth – so there is no opportunity for repair. Even in ideal conditions the butterfly will perish after some days, weeks or months, simply overcome by old age.

WHEN THINGS GO WRONG

Sometimes growth and development do not run smoothly, giving rise to malformed butterflies. These can include monstrosities with extra wings, legs where there should be antennae, or areas of coloured scales in the wrong place. But perhaps the most fascinating of all these abnormalities are 'halved gynandromorphs', which display both male and female characteristics. These unfortunate creatures are themselves infertile, but they tell us something very important about how butterflies, and insects in general, develop from the fertilized egg.

LEFT An unnatural event recorded in the collection of the Natural History Museum, London. This autumn leaf, *Doleschallia hexophthalmos*, had three antennae.

In mammals, including humans, the sex of an animal is determined by the XX (female) and XY (male) system of sex chromosome inheritance. All normal egg cells carry just one X sex chromosome, but sperm cells are produced in equal numbers of two sorts, X-bearing and Y-bearing. The genetic sex of a baby is determined simply by which sort of sperm fertilizes the egg: a Y-bearing sperm gives an XY (male) embryo and an X-bearing sperm gives an XX (female) embryo. Much the same happens in butterflies, except for the largely trivial detail that the roles are reversed: females are equivalent to XY, giving X and Y eggs, and all sperm are X-bearing. The crucial difference between mammal sex determination and insect sex determination depends on how the genetic difference is translated in the bodies that will grow from the fertilized eggs. In mammals only the Y chromosome carries instructions for making a protein required to form testes, which – very early in development – produce the hormone testosterone. This hormone perfuses the whole of the developing embryo by way of the bloodstream, leading to male development of the entire body. Without testosterone at this stage the embryo will develop as a female. In insects there are no developmental sex hormones of this type. Instead, each cell develops solely according to its own genetic sex and its precise location in the body.

A remarkable thing about insect development is that, of the two cells that are formed by the very first division of the fertilized egg, one goes on to form the clone of cells that will form the whole of one side of the body, while the other similarly forms the whole of the other side. No other symmetry is ever seen (e.g. top/bottom, or front/back). It also happens that the Y chromosome carries

BELOW This mixed-up mountain fritillary, *Boloria napaea,* is all male down the left-hand side, and all female on the right: a rare halved gynandromorph.

very little information. The sex of a cell is determined by the relative 'dose' of X chromosomes: in butterflies and moths, two X chromosomes in a cell make it develop as a male, one X (with or without a Y) as a female. In XX individuals otherwise destined to be normal males, very occasionally at that first division of the fertilized egg one of the X chromosomes gets lost. When this happens, the cells of whichever side received only one X then all develop as if they were female, while on the other side (XX) they all develop as male. This gives rise to the extraordinary halved gynandromorphs: butterflies that have literally 'split personalities', being all male on one side and all female on the other.

When this happens in a species that has a striking difference between the two sexes, the effect can be bizarre. Even the characteristic size, colour and shape differences are produced by the two halves. The separation extends right through the body, which struggles to make male sex organs on one side join up with female sex organs on the other. Almost needless to say, such malformed individuals are infertile.

Abnormalities can be induced by external factors, including radiation, chemicals (e.g. pesticides) and unusual temperatures. And, like ourselves, insects also suffer from diseases that can have a big impact during growth – including those caused by nematodes, fungi, viruses, bacteria and various micro-organisms. Much publicity has recently focussed on the devastating diseases of honey bees.

Ophryocystis elektroscirrha is a micro-organism that affects the monarch butterfly and its close relatives. Discovered in 1966, 'Oe' has unfortunately become a familiar problem to those who raise monarchs for interest, education, research – or commercially. Oe spores on the larval foodplant are ingested by monarch caterpillars. The parasites develop under the caterpillar skin, and multiply. During the pupal stage, new spores are formed around the scales on the head, thorax, wings and, especially, the abdomen. Although lightly infested monarchs appear to complete development normally, heavily infected individuals often have difficulty in completing adult emergence, and become deformed in the process. High parasite numbers, even in cases where eclosion is more or less normal, decrease adult survival and body mass. Much research is now being carried out on this particular butterfly disease – but not only is there much to learn about Oe, the general field of butterfly parasitology remains poorly known or understood. It is certain that many more butterfly diseases await discovery – and will thus add to our understanding of when and how things can go wrong.

ABOVE **Following the 2011 Fukushima nuclear disaster in Japan, a group of Japanese entomologists led by Atsuki Hiyama and Joji Otaki demonstrated induced and heritable deformities in populations of the pale grass blue,** *Zizeeria maha*, **at distances up to 80 km (49 miles) or more from the reactor site. The red arrows point to some of the deformities they found.**

CHAPTER 2
Mating

COURTSHIP, THE EXACT PATTERN AND DETAILS of which differ in almost every genus and species, is the core of a typical three-stage mating process: mate location, courtship and copulation. Would-be mates must first find each other, then the female must be satisfied that the male is suitable. If she accepts, then copulation and insemination ensue, after which the female must start laying her eggs. Meanwhile the male will seek further mates whenever the opportunity arises.

FINDING A MATE

In a vast savannah there may be few adults of any particular butterfly species, so the possibility of a couple meeting by chance is very low. 'Hill-topping' is an efficient strategy seen in some butterflies to overcome the location problem. Climb up a small hill that stands proud above the surrounding plain and you will often see male butterflies flying around the summit. As you approach they may dash out to check if you are a rival or a potential mate.

If you sit quietly and look down the hill, you may be lucky enough to see a fresh arrival flying up towards you. Resident males will fly out to investigate the newcomer. The encounter will be short if the two are different species. But if they are the same, then the reaction will be very different, and will depend on the sex of the newcomer. If it is male, there will be an aerial battle, in which the resident will usually succeed in defending his patch. When the intruder signals 'I give up', there will be headlong pursuit downhill until some critical point is reached, perhaps 50 m (150 ft) or so below the summit. There the resident will stop and swiftly return to the top, while the defeated male will fly on, in search of an unoccupied hill.

OPPOSITE **The great tree nymph or rice paper butterfly,** *Idea leuconoe*, **is now a common sight in butterfly houses. Like all milkweed butterflies, the males of this Southeast Asian species have special scent organs used during courtship. A male tree nymph will hover above a settled female for minutes on end, apparently wafting his scent over her in the hope of acceptance.**

BELOW *Coeliades pisistratus,* the two-pip policeman, photographed at Wli, Ghana. Like many skippers, the males fly very actively and fast, and frequently investigate the tops of small hills and other prominent places, where they are always ready to pursue a potential rival – or, if lucky, a passing female.

RIGHT **A mixed assembly of butterflies – whites, sulphurs, swallowtails and nymphalids – with a solitary day-flying moth (far right), on muddy rocks by a South American river. Occasionally a female may fly into such a group of males, in the hope of eliciting sex.**

When a resident male encounters a female of his own species, however, he will switch to courtship behaviour. This is often prolonged, as the female seeks to establish not only that the male is a member of her own species, but also that he would be a good or 'suitable' sire for her offspring. Very often the female behaves coyly, flying off without signalling acceptance. To prove his worth the male must keep up; if she flies down the slope, he will not stop at the 'critical point', but will fly on and on until he either succeeds and is accepted, or she rejects him. Either way, later on he must seek out another hilltop and hope to establish himself as the resident. Males sometimes occupy a hilltop for a week or more, waiting for a female to come along.

Hilltops are an effective way of getting the sexes together. Males seek out such places, and females that need to mate do likewise. Those that are already mated and are laying eggs avoid them. Other features in the landscape can act as focal points, such as tall trees, clearings in a forest, or even a sunspot in a wood. Such vantage points are occupied by males that perch and wait for females to come along, dashing out to investigate anything that moves. As in hill-topping, the males aggressively defend their chosen station against rivals.

An alternative male tactic is patrolling. Males fly continuously through a large area, along tracks, woodland edges or following streams, on the look-out for females. Visiting larval host plants, where pupae are likely to be more numerous, is a key part of this strategy, as encountering virgin females is the best potential pay-off. If a patrolling male encounters another male, the interaction is typically vigorous but brief. They whirl around each other for a few moments as they establish that they are not prospective mates; once satisfied, they break off and continue on their way. But if they encounter a female or see one resting, they will close in. Rarely, it seems, the roles may be reversed. Males of many species sometimes congregate in large numbers on mud and wet sand, and females have been seen to fly into these groups, apparently to persuade one or other of them to initiate courtship.

COURTSHIP

The Papuan birdwing, *Troides oblongomaculatus*, is impressive in courtship as well as in size. If a male approaches a flying female from behind, he will overtake her from below (sometimes called 'undertaking'). As soon as he is just in front he floats up, brushing his special white hindwing scale patches against the head of the advancing female. As he does so, some fluffy filaments break off and stick to her antennae. These little pieces of scale are thought to carry pheromones – chemical messengers that are important, or even essential, for successful courtship. The female carries on flying in a straight line, while the male, now above her, slows a little and falls behind. Resuming his pursuit, he undertakes again, flying up to brush her antennae a second time. This looping sequence is repeated several times.

If the female has already been mated, she will almost certainly refuse the male. She does this by dropping to the ground, with wings spread out and her abdomen thrust downwards. He will soon get the message. If, on the other hand, the female settles on a bush, the male will continue his courtship by hovering directly above her, displaying his brilliant yellow hindwings and the white scale patches. At last she may finally accept him and then the male will settle alongside and copulate. They will stay together for up to five hours.

In alfalfa fields in North America two species of sulphur or clouded-yellow butterflies, *Colias eurytheme* and *C. philodice*, can be so numerous that courtship behaviour takes on the role of ensuring that the butterflies mate with their own species. This is crucial, because if they make a mistake hybrid offspring will be infertile. To our eyes the two species look very similar, but if you photograph

LEFT **Three male *Troides hypolitus* chase a female, photographed on the island of Ambon, Indonesia. The courtship of this species has not been recorded in detail, but it is almost certainly similar to that of the Papuan birdwing.**

RIGHT AND BELOW Both *Colias philodice* (top row) and *C. eurytheme* (second row) have yellow or white female forms (middle and right columns). Photographed in ultraviolet light (lower two rows), all the females as well as the male *C. philodice* appear dark, but the yellow areas of the male *C. eurytheme* reflect UV brilliantly.

them in ultraviolet (UV) light you see a striking difference. The wings of male *C. eurytheme* reflect UV very strongly, whereas their females, and both sexes of *C. philodice*, do not. The males of both species search for females visually. If they encounter a butterfly that reflects UV, like a male *C. eurytheme*, the males of both species are repelled. If a female *C. eurytheme* is approached by a male, however, she will only allow courtship to go ahead if her suitor is UV-reflectant. In this way, she can avoid the unwanted attentions of *C. philodice* males. But what of female *C. philodice* – do UV-reflecting males repel them? Apparently not: they seem indifferent to male colour. To be acceptable, male *C. philodice* must have the correct smell, or pheromone 'bouquet', a mixture of volatile chemicals that includes three particular compounds called esters. It turns out that smell is also vital to the *C. eurytheme* females, but their males must look correct as well if courtship is to proceed.

Apart from making sure her mate is of the right species, females also use courtship to decide whether a male is suitable to mate with. In many butterflies the female sets the male a task of following her in a complex ritual. In some cases this consists of a headlong, whirling aerial dance. By keeping up and following the correct sequence, rather like following the right steps in ballroom dancing, the male can demonstrate that he is strong and well coordinated. Individual males unable to do this may be old and less fertile, or constitutionally weak because of some genetic defect. Pheromones also play a role here. The smell of the male must be correct for the species, but the mix of chemicals in the male bouquet is often complex, and individual variation may indicate greater or lesser suitability as a mate.

The courtship display of the European grayling, *Hipparchia semele*, has been closely observed. These butterflies live in open areas, where they often form dense colonies. Once a male and receptive female have found each other, courtship starts with the pair standing face to face. The male then opens his wings and performs a sort of 'bowing' movement. He then leans forward and takes the female's antennae (her main organs of smell) between his wings and closes them together. Next he slowly tips back, dragging the central area of his forewings along the length of her antennae. Close examination of the male reveals that he has special, apparently glandular scales in this area, not found in the females, and that this movement brings his main source of pheromones into close contact with receptors on her antennae. The bowing process is repeated several times, after which the male turns round to stand alongside the female and attempts to mate. If he has performed well she may accept him, but if not, or if she is uncertain, she will suddenly fly off. The male will usually pursue her and try to start the courtship process again.

LEFT Courtship for small tortoiseshells can be an exhausting business. Pursuit of a female can last up to three hours. In the ground phase seen here, a male is closing on a female from behind, in the hope of consummation.

RIGHT The hindwing pouch of a male blue tiger, *Tirumala*, seen from below. The pouch is filled with tiny particles of 'love dust' (close-up above) which, after transfer to the butterfly's abdominal hair-pencil (as seen here) are used in courtship.

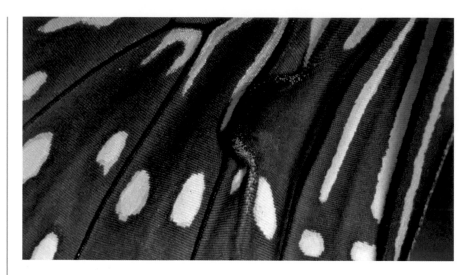

PRE-COURTSHIP RITUALS

The males of some butterflies have to prepare themselves in advance of the complex business of courting. When a male of the rare South American genus *Antirrhea* emerges from its pupa, he rubs specialized scales from a patch on his forewings onto brushes or 'hair-pencils' at the tip of his abdomen. It is thought that the male later transfers these minute scales to the antennae of the female during courtship, as with the white fluff of the male birdwings, presumably carrying with them one or more chemical messages. A similar process occurs in the blue tiger milkweed butterflies (genus *Tirumala*), whereby the male inserts an abdominal hair-pencil into a deep pouch in each of his hindwings. Inside the pouches, special scales produce thousands of tiny studded particles, and it is these that are used as little sachets to convey chemical messages to the female.

Milkweed butterflies such as *Tirumala* do more than this to prepare for courtship, however. Day after day the males seek out particular plants, such as certain borages and daisies, from which they obtain a poisonous alkaloid (see p.70). They alter some of the chemical to form a very potent and essential part of their pheromone bouquet. Without this as part of their aroma, the females would be almost certain to reject them.

COPULATION

Having at last gained acceptance, the male butterfly can then inseminate the female. To do this he must insert his phallus (aedeagus or penis) into the vulva (vestibulum) of the female. To do this successfully he must engage her abdomen with a special locking device. At the tip of his abdomen there is a fixed, hook-like structure (the uncus) with two lateral arms (valves) that close together to give a strong grip. Once firmly locked together, the tip of the phallus is then exerted

LEFT A pairing of common blues, *Polyommatus icarus*, in a Kent nature reserve. The male is on the right. On the upperside of his forewings he has scattered 'androconial' scales, the source of scent in many male lycaenid butterflies.

between the valves to reach into the vestibulum, and then lengthened by the end part (the vesica) popping out, like the finger of a rubber glove, to enter the vagina proper. The male transfers his sperm to the female via this tube. It is a slow process, usually taking at least an hour, sometimes much longer, depending on the species.

The seminal fluid is contained within a special packet called a spermatophore. This packet is transferred to a reservoir within the female (the bursa copulatrix), where it remains viable and capable of fertilizing eggs for many days. To release the contents of the spermatophore, the female must rupture it; this is usually assisted by special teeth on the inside of the bursa. The eggs move along a completely separate duct. Just before an egg is laid, it receives a little quantity of sperm that is moved, under the control of the female, from the bursa to the egg duct via a connecting tube. The sperm enters the egg through the micropyle.

Over the last 30 years evidence has accumulated that in most butterflies far more than just sperm is transferred with the spermatophore. In many cases the packet also contains proteins, carbohydrates and fats that the female can use to make more

BELOW A coupling of the common rose, *Pachliopta aristolochiae*, photographed in southern India. The male swallowtail (below) probably found his mate as she was emerging from the pupa, as her wings are not yet fully expanded.

eggs, and so increase her fecundity (and the reproductive success of her partner). The spermatophore sometimes makes up as much as 10% of the male's total body weight, so this male contribution can be really significant.

In the case of the milkweed butterflies, such as *Tirumala* and its relatives, even more is passed on. The spermatophore contains some of the poisonous alkaloids that the males collect so avidly. The male's contribution is added to whatever alkaloid the female is able to obtain, and is passed on to the eggs to protect them against various spiders and insects that might otherwise devour them.

HOW MANY TIMES DOES A BUTTERFLY MATE?

As with so many such questions about butterflies, it all depends on sex and species. Females of some species, notably those that lay all their eggs within a day or two after mating, usually mate only once. By contrast, those that produce eggs over a long period, perhaps weeks or even a few months, will mate several times. This may be to replenish their store of sperm, to get more viable sperm, or to gain additional nutrients that will help make more eggs. The females of some milkweed butterflies seem to mate frequently, perhaps to obtain continual supplies of alkaloids, both for the protection of their eggs and perhaps their own survival, as the alkaloids are highly unpalatable to many predators, including birds.

The males of most species mate as many times as they can, sometimes twice in one day. On the one hand, some are lucky, mating several times before they die. On the other hand, quite a few unsuccessful males die as virgins. This seems to contrast with females, most of which mate at least once. This asymmetry reflects the potential for selection by females and their ability in most species to refuse the advances of unwanted or unattractive males. In many cases we can see that the females have a clear mate-refusal signal, such as the vertically upturned abdomen used by white butterflies. As the unrequited male hovers and fusses around the unimpressed female, to the uninitiated this may look like a 'come-on' signal. But this position makes copulation almost impossible, and if you watch you will see the male sooner or later lose interest and fly off in hope of finding a more willing partner.

One of several reasons for rejecting a male is that he is old. As the males of most species age, they produce smaller and smaller spermatophores. So the reward for mating with an old male may simply be too low.

Wing wear and other cues may give away the male's relative age. However, in case you think that females have all the aces in the mating game, the males of some species have a way of protecting their investment. As copulation is completed, the male plugs the female's vestibulum so that no other male can gain access. Although this plug may eventually dissolve or fall out, so that long-lived females of these species can benefit from a second mating, this usually means that most or all of the sperm from the first male will be used by the female, rather than just a

little of it. When a female mates again, the sperm of the latest mate usually replaces any sperm left from a previous mating. The apollo butterflies (genus *Parnassius*) exhibit this mating-plug phenomenon very well, putting in place a 'sphragis' that can be large and characteristic of particular species.

In at least one group of butterflies, a subgroup of the genus *Heliconius*, the phenomenon of 'pupal rape' occurs. Males are able to locate female pupae of their own species that are just about to hatch. They scramble to mount the pupa, aiming to be the first to insert their abdomen inside and copulate with the female just as the case of the pupa splits. So, unlike most butterflies, females of these species have no control over who will mate with them; for these seemingly ruthless males, fancy courtship is a thing of the past.

However, it should not be imagined that mating behaviour in different species, however characteristic, is always pursued in the same way without variation. The males of some butterflies, such as the speckled wood, can alternate between perching or patrolling tactics, depending on circumstances. In the New Guinea yellow birdwing, the impressive courtship already described is not followed if a male finds a female pupa that is about to hatch. Instead he will simply hover above the pupa, and mating will take place soon after emergence, even before the wings of the female have fully hardened.

ABOVE Mating *Parnassius apollo*. The female is to the right. To prevent later suitors taking advantage, the males of many apollo butterflies produce a sort of chastity belt ('sphragis'), to seal the female's genital opening once the spermatophore has been transferred.

CHAPTER 3
Laying

ALTHOUGH MOST BUTTERFLIES OFFER no parental care whatsoever, the females go to great lengths to ensure that their eggs have the best possible chance of survival. They search for the right food plant and even calculate whether there will be enough food available. But there is at least one species in which the female butterfly seems to make a much greater commitment to her offspring.

THE CURIOUS CASE OF THE MALAYAN EGGFLY

In April 1979, Gweneth and Bernard Johnston made a remarkable discovery. They were at Catmon, on the island of Leyte in the Philippines, looking for early stages of a butterfly called the Malayan eggfly, *Hypolimnas anomala*. At first they and their Filipino colleague, Manuel Medicielo, had no luck at all, despite knowing the correct food plant for the caterpillars and going to an area where the plant and the butterfly were both common. After searching nearly all day without success, a young boy who was helping them suddenly produced a batch of eggs on a leaf – with a female eggfly standing astride them.

Delighted, they took the detached leaf, eggs and butterfly back to their temporary laboratory. To their surprise, the butterfly made no attempt to fly off, remaining rigidly in place, almost as if standing to attention. Even the flash from a camera had no effect. If touched, the eggfly snapped her wings open and shut, but remained at her post, hour after hour. The next day, the team discovered that finding batches of eggs among the trees was easy after all – they just had to look out for female eggflies beneath the leaves. Wherever they found one, it was standing over a big cluster of eggs.

Subsequent studies on the Malayan eggfly, and its very close relative *H. antilope*, have demonstrated that this behaviour, probably unique among the butterflies, is an effective form of

OPPOSITE **A female atala butterfly, *Eumaeus atala*, in the process of laying 15 of her eggs at Boca Raton, Florida. The caterpillars of these butterflies can only feed on various cycads, a group of non-flowering plants related to conifers.**

BELOW **A female *Hypolimnas anomala* photographed in the Philippines. She will stand guard over her batch of eggs, laid on the underside of a leaf of a tree belonging to the stinging-nettle family, for over five days.**

parental care. Female *H. anomala* lay unusually large batches of eggs: typically about 500, although a batch of more than 1,400 has been recorded. By standing guard for up to five or more days until the eggs hatch, and a further one or two days while the tiny first stage larvae begin to feed, the mother can increase the rate of survival of her eggs by up to 50%. She achieves this mainly by driving off small ants that suck out the contents of the eggs, although she is relatively powerless against larger ants that can carry the eggs away, or minute parasitic wasps.

Interestingly, some of the guarding females can be scared off. This possibly depends on the degree of investment in a clutch. Small batches, below 100 or so, have very poor survivorship, even if guarded. Females that more readily abandon their eggs may be those that still have lots of eggs to lay, but this has not been established. In the case of those that are steadfast and immovable, however, it seems they are prepared to sacrifice their own lives in the hope of benefiting their young.

If the Malayan eggfly is unique in its egg-guarding behaviour, does it mean that all other butterflies simply abandon their eggs? The answer is 'yes and no'. So far as we know all other sorts of butterflies, once they have laid their eggs, do fly off and leave them to their fate. However, in most cases they actually take a great deal of care over where and how they place them.

LOCATION, LOCATION, LOCATION!

Names such as holly blue, spicebush swallowtail, milkweed and caper white tell you two fundamental things about the biology of butterflies: most of their caterpillars feed on plants, and most can only feed on a few particular sorts. The spectacular birdwings never feed on anything other than vines belonging

RIGHT An Australian jezebel, *Delias harpalyce*, laying a string of eggs on drooping mistletoe at Morwell National Park, Victoria. Note the fat abdomen, still crammed with eggs. The females of many of the 250 or so species of these fascinating pierid butterflies must find exactly the right sort of mistletoe if their larvae are to survive.

to the family Aristolochiaceae, while the gaudy 'jezebels' of the genus *Delias* focus on mistletoes (family Loranthaceae) and their close relatives. In the mountains of New Guinea both sorts of butterfly abound. But move a *Delias* caterpillar onto a vine on which a birdwing larva is feeding, and swap the birdwing caterpillar onto the mistletoe, and both would starve to death rather than eat the strange plant. In reality they would wander off in search of the correct food plant, spurning all else until they found what they needed. But given the incredible diversity of plants in the tropics, a small creeping caterpillar generally has little chance. Even though surrounded by greenery, if separated from the plant on which it hatched, it will nearly always end up starving to death, following a desperate and almost hopeless search. So the first rule of parental care for most butterflies is 'find the right plant'. Lay on the wrong sort and your offspring will almost certainly die.

Watch a female of the American bordered patch butterfly, *Chlosyne lacinia*, and you may see her seeking out one of the several species of daisies on which she can lay her eggs. She picks out a particular plant, lands on a leaf and then begins to scrape at it with a drumming motion of her spiny forelegs. Special receptors, located right alongside the spines, pick up chemical cues as to the species of plant and whether it is suitable for her eggs. If satisfied she may lay a number of eggs; if not, she will fly off in search of the correct species or a better individual to host her precious young. The females of most butterflies follow an equivalent routine, whether the food plant is a tiny herb or a huge tree. Once the butterfly accepts a particular plant, she will usually select a suitable leaf and curl her abdomen around its margin to lay on the underside. As an egg passes down the oviduct it is fertilized, and then smeared with a special 'glue' that sticks the egg to the leaf surface.

LEFT **As well as consuming butterfly eggs, parasitic wasps prey on their larvae. The mother of this handsome wasp, a species of the genus** *Mesochorus*, **laid her egg in a dwarf crow butterfly caterpillar, now eaten out and mummified. The wasp's exit hole is clearly visible. However, this particular wasp did not eat the actual butterfly larva – it ate the larva of another parasitic wasp (family Braconidae) that had already developed inside!** *Mesochorus*, **which belongs to the wasp family Ichneumonidae, is what is known as a hyperparasitoid.**

ONE EGG AT A TIME, OR MANY?

Having found the correct plant, perhaps after a long search, does the female simply lay all her eggs in one go? If you follow a female *Heliconius erato* butterfly intent on egg laying, you may see her examine a passion flower plant for several minutes before laying just one egg and flying on. By contrast, *Heliconius* species such as *H. xanthocles* and *H. sara* lay their eggs in batches, and large batches can be built up by several *H. xanthocles* choosing to lay at the same spot.

The fact that two closely related *Heliconius* butterflies lay their eggs in different ways suggests that there may be advantages and disadvantages to these two basic egg-laying strategies. Probably most butterflies lay their eggs in ones, twos or threes, and then move on to another plant. This way they avoid the dangers of 'putting all their eggs in one basket'. Those that feed on small herbs avoid the risks of starvation for the developing caterpillars if they spread their eggs between many plants. Also, laying in ones and twos can provide less of a target for predators. For example, most species are afflicted by tiny parasitic wasps that lay their eggs inside the butterfly's eggs. The minute grubs that hatch out devour the butterfly eggs from within. A single female wasp could destroy one large batch of eggs, but by laying solitary, widely distributed eggs, the female butterfly ensures that at least some will survive. The goal is to ensure the survival of the highest number of eggs possible.

On the other hand, laying eggs in large batches, like the Malayan eggfly or *H. xanthocles*, can have real benefits too. Once a female has found a suitable plant it might be better to place many eggs there, rather than run the risk of a long search for other suitable plants on which to lay eggs singly or in small numbers. Suitable plants may be few and far between, and every flight makes the female conspicuous to potential enemies. Eggs laid together can also offer safety in numbers. It seems that in some circumstances those on the outside of a clump can save the eggs at the centre from a parasitic wasp attack, or from drying out. If the eggs contain some form of chemical protection against predators, then clustering may make the defence far more effective, because it will increase the dose. The larvae that hatch out are often chemically defended too, and such larvae usually feed together in a group, surviving and developing better as a result.

Egg laying in batches can take several forms. Apart from making a single sheet of eggs, in which each egg may be in contact with its neighbours, or a distinct space apart,

BELOW Two *Heliconius charithonia* laying a joint batch of eggs. In the wild this species usually lays a few eggs at a time, but in captivity females may switch to batch laying, and even lay their eggs together – underlining the ability of butterflies to be adaptable (see box opposite).

LEFT *Heliconius erato* lays only one or two eggs at a time. Unlike *H. xanthocles* (see text), it has never been observed to lay batches of eggs, or lay gregariously with other females - either in the wild, or in captivity.

LEFT *Heliconius sara* is a batch layer. This behaviour has been observed in the wild and in the laboratory. Moreover, two females have been observed to lay gregariously, to form a collective batch - again, both in nature and in captivity. But unlike *H. xanthocles*, in this species, gregarious egg laying involving many individuals has not been seen.

CHANGING STRATEGIES

In 1949 Erik and Astrid Nielsen made a study of the great southern white, *Ascia monuste*. In Florida they found that the larvae of this butterfly feed on a number of plants, including small, narrow-leafed species such as saltwort, *Batis*, and peppergrass, *Lepidium*, and more substantial, broad-leafed plants such as nasturtium, *Tropaeolum*, and spider flower, *Cleome*. On encountering a suitable narrow-leafed host plant the butterflies laid single eggs, but on coming across a suitable broad-leafed host they would lay in clusters of up to 50. The Nielsens also noticed that the larvae from clustered eggs stayed in communal webs until their third instar. Although it is not clear whether an individual female can switch from one strategy to the other, these discoveries suggest that in some cases it may simply be the size of the host plant, and its ability to sustain a few larvae or many, that determines the pattern of egg laying. It is also evident that a female butterfly can assess the capacity of a plant to sustain her offspring – and how many eggs to lay on a particular plant.

ABOVE The map is one of
Europe's most interesting
butterflies. Late spring European
maps are mainly orange above,
whereas late summer individuals
are largely black. Here a spring
generation female is laying the
characteristic chains or strings
of eggs on nettle.

some butterflies lay in masses that are three or four layers deep. The most striking alternative involves pendant chains of eggs – seen, for example, in the European map butterfly, *Araschnia levana*. The female lays several chains 10–12 eggs long, which can be seen hanging together below a leaf, like strings of beads.

BUTTERFLY ARITHMETIC, FALSE EGGS AND LEAF DIVERSITY

This sensitivity to density, and in particular the capacity of females to assess the egg load on a particular plant, points to something akin to mathematical ability in butterflies. Miriam Rothschild, in collaboration with Louis M. Schoonhoven, conducted laboratory experiments on the egg-laying behaviour of the large white, *Pieris brassicae*. They found that this familiar European insect used a variety of cues to assess the load of eggs and larvae already present on a cabbage plant, including chemicals released by hatched eggs, crushed eggs or feeding caterpillars, as well as potential visual markers such as the number of eggs. In their experiments, if the density of eggs or feeding larvae exceeded a certain level, the large white females would not lay, but below some critical value they would. Plastic egg dummies and spots of yellow paint were effective in eliciting the response, but model caterpillars were not.

Such competition for places to lay an egg can have some odd consequences. Perhaps the most notable occurs amongst the passionflower butterflies. Although passion vines can grow fairly large, the new, unfolding leaves are crucial for the young *Heliconius* caterpillars. On coming across a newly hatched larva, an older caterpillar will eat it if there is insufficient new plant growth to satisfy them both. Most *Heliconius* females lay only one or a few eggs per day, so it is crucial that they do not put any where they are likely to die. Before laying a single egg a female spends minutes fluttering around a plant to assess its suitability. This includes, so it seems, 'counting' how many other eggs and young caterpillars are already present. If there are too many in relation to the number of young vine leaves, she will not lay.

In the early 1970s the great *Heliconius* specialist Larry Gilbert discovered that the leaves of some passion vines have little yellow 'warts' on them that look just like butterfly eggs. Butterflies encountering vines with these egg-dummies apparently treat them as the real thing – and abandon what is in fact a perfectly good place to lay. One up to the plants! A similar phenomenon has been found to affect the North American spring white butterfly, *Pontia sisymbrii*, which lays its red eggs on the crucifer genus *Streptanthus*. Little red warts on certain *Streptanthus* species deter *P. sisymbrii* females from laying. Gilbert also

LEFT The two yellow warts at the base of this passion-vine leaf are part of the plant. The *Heliconius* egg to the left was placed there by Larry Gilbert, for comparison. In nature the dummy eggs deter the butterflies from laying.

demonstrated that *Heliconius* females conduct their initial search for passion vines visually, looking for leaves of the right shape. He recognized that a remarkable diversity of leaf shape occurs in this group of plants, and suggested that by continually evolving new shapes, the plants make themselves more difficult for *Heliconius* butterflies to find.

LAYING EGGS IN STRANGE PLACES

Watch the egg-laying behaviour of the Asian 'brownie' *Allotinus major*, and you may see something rather strange. As she flutters from plant to plant, she only seems interested in those with clusters of a particular plant-sucking bug, which may or may not be attended by numerous ants. Plants without the bugs are spurned. The brownie, a member of the 'blue' family that has many ecological interactions with ants, is one of those exceptions in the butterfly world – it is a predator. Its caterpillars feed on young membracids, a group of insects related to aphids. These little bugs are often attended by ants, which, in return for offering some protection, obtain nourishing exudations from them. The brownie is able to overcome the normal aggressive reaction of ants to intruders and will settle near the bugs to lay her deadly crop of eggs. Although the ability to get so close to the bugs almost certainly depends on chemical trickery to fool the guardian ants, the butterflies have physical specializations too. Like the closely related *Miletus* (see p. 73), *Allotinus* butterflies have long legs, which may act like stilts so that they can stand over their prey and the attendant ants.

The larvae of the remarkable moth butterfly, *Liphyra brassolis*, eat nothing but the grubs of the weaver ant, *Oecophylla smaragdina*. Weavers live in trees, sewing leaves together to make a series of large nests, which they defend with terrible ferocity. The female butterfly flies rapidly around an infested tree to check its suitability, before landing on the underside of a branch close to the trunk, or even on the trunk itself, to lay a single egg. She may repeat this a few times, to leave two or three eggs, sometimes as many as six, on a tree. How do the small first-instar larvae get into a nest? No one seems to have observed this (being close to a weaver nest is not much fun!), but it is apparent that older larvae can move between nests on the same tree and first-instar larvae have been found within weaver ant nests.

RIGHT The larvae of some lycaenid butterflies, such as this second instar *Allotinus substrigosus* from Malaysia, eat aphids and similar sapsucking bugs tended by ants. The lycaenids usually avoid being attacked by mimicking the ants' scents – and by offering nutritious rewards.

RIGHT Species of the South African genus *Torynesis* only occur where their host grass (*Danthonia*) grows. This female 'mintha widow' lays her eggs with care. Other satyrines are less choosy, and some species just scatter their eggs as they fly.

OTHER OVIPOSITION STRATEGIES

So far we have seen how most butterflies oviposit – lay their eggs – exactly where the caterpillars need to be to get not only the right sort, but also an adequate amount of the particular food they need to develop. This can include what at first sight might seem like crazy places, among plant bugs or even next to the nests of ferocious ants. But oviposition sometimes has to take into account other ecological factors as well – including the seasons of the year.

Watch a female of the lovely European silver-washed fritillary, *Argynnis paphia*, egg-laying. Her caterpillars can only survive and grow by feeding on violets, *Viola* sp., but she chooses instead to lay her eggs on a tree trunk – typically an oak, placing her eggs a metre or more (3 ft) above the ground. In this species, rather remarkably, it is the first larval instar that has the 'responsibility' of surviving the winter. The eggs hatch in autumn and the minute first instar larvae hibernate without feeding. Nestled securely in a north-facing, moss covered bark crevice they are far better protected than being down on the ground, among the tiny violets. But the mother will always choose a tree with violets close to its base – usually not more than 2 m away. In the spring the larvae are then faced with a hazardous trek down the tree trunk, to find their food at the bottom – to scale, a bit like asking a newborn human baby to sleep for six months and then crawl up to three laps of an Olympic running track before having its first drink of milk!

BELOW A female silver-washed fritillary ('valesina' form) photographed in summer laying on an oak tree in a Sussex woodland. The tiny larvae hatch and overwinter without feeding, secure in the crevices of the tree bark. But what do they feed on in spring, after hibernation?

Some butterflies that feed on plants, notably those that eat grasses, operate a less precise egg-laying strategy. As grasses occur in extensive mats or swathes that spread over entire valleys or hillsides, the females of some 'brown' butterflies take advantage of this, merely fluttering across the sward, scattering their eggs as they go. This tactic is probably not as careless as it seems. Almost certainly the female butterflies instinctively appreciate subtle differences in the lie of the land, and release their eggs in the areas where the grass grows best, or is in the best location to provide the right climatic conditions for their caterpillars later in the season.

CHAPTER 4
Eating

CATERPILLARS BECOME NON-STOP EATING machines almost as soon as they emerge from their eggs, as many a dismayed gardener will testify. Adult butterflies also need to feed in order to have the energy to fly, find a mate and reproduce. However, as we have seen at the end of the previous chapter, not all caterpillars eat green food – nor do all adult butterflies sip nectar.

JAWS

Put some monarch caterpillars, *Danaus plexippus*, in a jar with their favourite milkweed leaves to feed on, stretch a paper cover over the top, and listen. You will hear a steady munching sound, like someone biting into a crisp apple, going on and on and on, hour after hour, day after day. Welcome to the world of the caterpillar: a world of eating, a world of jaws.

The jaws of a caterpillar cut and chew with a side-to-side motion, unlike the up-and-down movement of mammal jaws. The caterpillar's jaws require large muscles to power them, and these fill most of its head. Between the jaws is a little tube or spinneret that produces silk, a vital accessory for larval life, and a pair of articulated lobes that assist with manipulation. On either side of the paired jaws is a short antenna, an organ of scent and touch, and a horseshoe-shaped array of six tiny lenses, making a simple eye that can probably do little more than tell the difference between light and dark.

So equipped, the caterpillar must ingest all the food needed to grow into an adult butterfly. After it has eaten its eggshell, if it needs to do that, the first-instar larva must start on the green leaves chosen for it by its mother. Often the thickness of the whole leaf is too much for its jaws at this stage, and so it eats away at just one of the outer layers of the leaf, often working characteristic little circles out of the leaf surface. Once into the second instar, the larva will normally transfer its attentions to the leaf edge, by now having jaws and muscles large enough to cope with the whole thickness of a leaf.

OPPOSITE This studio 'portrait' shows a final instar caterpillar of *Papilio machaon*, the swallowtail, doing what it does best – eating. It is feeding on fennel, one of its favourite host plants.

BELOW Head of a *Caligo* caterpillar from below. The jaws are tightly shut, but their hardened black cutting edges are clearly visible, as well as the spinneret, antennae, simple eyes and first pair of legs.

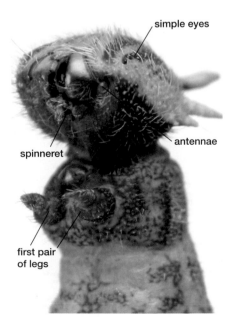
simple eyes
antennae
spinneret
first pair of legs

Leaves vary greatly in texture and quality: they may be hard, fibrous, fleshy or soft. And the jaws of caterpillars vary, depending on what they eat. The caterpillars of most butterflies feed on fibrous plants and generally have serrated jaws, like pairs of steak knives. Grasses are frequently hardened with silica, and caterpillars that feed on them often have smooth and apparently very hard jaws, like shears.

Food quality may also affect the muscle power needed and therefore the size of the head. Some years ago Phil DeVries discovered that the rare Latin American butterflies that belong to the genus *Antirrhea* feed exclusively on the old, mature leaves of palms. The leaves are incredibly tough, and this almost certainly explains the disproportionate size of the first-instar head, which makes the caterpillar look a bit like a tadpole: to get started on such tough stuff, these little caterpillars must have a fearsome bite.

WHAT PLANTS DO CATERPILLARS EAT?

There are about 300,000 different sorts of green plants, divided among some 300 families, and butterfly larvae use a very wide variety of these as food. Most individual species, however, specialise on just a few sorts. Monarch butterfly larvae have been recorded feeding on well over 40 different species, but nearly all belong to a single subfamily, the Asclepiadoideae or milkweeds, part of the dogbane family, Apocynaceae. The small white, *Pieris rapae*, scourge of cabbage growers in many regions of the world, has been recorded on perhaps 30 plant species, and nearly all of these are members of the cabbage family, Brassicaceae. A very informative exception, however, is nasturtium, *Tropaeolum*. This ornamental

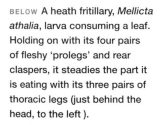
BELOW A heath fritillary, *Mellicta athalia*, larva consuming a leaf. Holding on with its four pairs of fleshy 'prolegs' and rear claspers, it steadies the part it is eating with its three pairs of thoracic legs (just behind the head, to the left).

plant is not closely related to cabbages, but it does have something in common with them: a hot, peppery taste. Both sorts of plant produce a very similar glucoside (a pungent chemical that is thought to protect the plants against a wide variety of potential enemies). This same substance may be what females of the specialised *Pieris* butterflies need to detect in a plant before they will lay on it.

PLANT DEFENCE TACTICS

The glucosides found in cabbages and nasturtiums, and other so called 'secondary' plant compounds, do not seem to be essential for plant growth but are thought to have evolved for defence. They are also believed to be a major factor in butterfly evolution. If a plant 'invents' a new secondary compound that is effective in deterring attack, then it may be able to proliferate and even diversify. Such plants then have huge potential as untapped resources for any herbivore that can solve the chemical challenge. Evolutionary biologists have imagined a type of biochemical arms race: each time a plant evolves a new chemical defence, although this gives a temporary advantage, the herbivores sooner or later respond by altering their own chemical ability to cope.

If one group of plants and one group of herbivores, say a genus of butterflies, were to get locked into a long-running evolutionary 'tit for tat', we could speak of this process as 'co-evolution'. For a while this was a very fashionable idea about the evolution of butterflies as a whole: butterfly and plant diversity were thought to have evolved together, giving a distinct pattern to what each sort of butterfly would and could eat. However, it now seems more realistic to assume that plant defensive chemistry has been a response to the whole range of herbivores, including locusts, bugs, beetles and wasps, in addition to moths and butterflies, not to mention molluscs, vertebrates and other plant eaters. If so, then in order to take advantage of the wide diversity of plants, many posing their own more or less unique chemical challenges, the butterflies have had to diversify too. According to this alternative idea, butterflies have not so much co-evolved with plants, but have tracked or 'radiated' on to them.

One of the most closely linked associations among butterflies and plants is that between the swallowtail tribe Troidini, to which the birdwings belong, and Dutchman's pipe vines, which belong to the plant family Aristolochiaceae. No larva of the 130 species of troidine swallowtails has ever reliably been recorded feeding on any other plant group. The vines are characterized by a distasteful secondary compound called aristolochic acid, which the troidine caterpillars are able to cope with. The troidines may be able to store or process the aristolochic acid to their own advantage, as these swallowtails are relatively immune from attack by predators, and have bright colours to advertise this. Among other butterflies, just a few members of the unrelated Lycaenidae family eat the vines, feeding on the flowers and seedpods rather than the leaves and stems, which the troidines use.

ABOVE The poisonous pyrrolizidine alkaloids found in ragwort occur in several other plants, including pink-flowering *Eupatorium.* Adult milkweed butterflies, like these chocolate tigers, *Parantica sita,* seek out these alkaloids for their own defence – and to make essential sex pheromones.

RIGHT This strikingly patterned chocolate tiger caterpillar can only feed on certain species of milkweeds (family Apocynaceae). Once mated, the adult females must find one of these plants on which to lay, if her young are to grow up successfully.

In some contrast the Charaxini, a group of about 200 species of nymphalids to which the handsome African *Charaxes* belong, utilise over 30 plant families, mostly various sorts of trees. Although none of these butterfly species is ever likely to be recorded on all 30 plant families, *C. jasius*, for example, has so far been found on at least six. As butterflies go, the larvae of Charaxini are catholic in their eating habits. These butterflies apparently lack chemical defences; instead, they rely on toughness and exceptional flight speed to escape from their enemies.

Do plants have any other defences against caterpillars, other than poisonous or distasteful compounds? Grasses are defended mechanically, by silica-hardened leaves. Some plants produce latex, a sticky sap that exudes from cut leaves and stems. When a larva bites into a latex-bearing leaf, its jaws become enmeshed in a fast-drying rubber that can literally 'gum it up' forever. Species that feed on such plants, like the milkweed butterflies, have overcome this defence. They cut a notch in a leaf petiole and wait until most of the latex has dripped out, before starting to feed. Passion vines, which despite their chemical defences are badly afflicted by *Heliconius* butterflies, try a variety of tricks to deter their attacks, including the dummy eggs mentioned earlier (p.37). One of these vines, however, seems to have won the arms race by purely mechanical means. If a *Heliconius* female lays an egg on *Passiflora adenopoda*, the result is invariably fatal. When it hatches, the movement of the young larva stimulates sensitive hooks all over the leaf surface, which contract and literally impale the little caterpillar in a death-grip. Older *Heliconius* larvae, if artificially transferred onto this plant, can feed with impunity due to their size and strength – but in nature this does not happen.

LEFT These African *Charaxes zoolina* are feeding on sap. The larvae of this particular *Charaxes* are only known to feed on thorns, such as *Acacia*, but other members of the genus use a wide variety of bushes and trees as hosts.

ALTERNATIVES TO LEAVES

Not all butterfly larvae feed on leaves. Many specialize on other plant parts, including buds, developing flowers and seeds. For butterflies of the blue family, Lycaenidae, these are often very attractive as food because of their high nitrogen content. Many blues have close ecological relationships with ants, which they appease with sugary and amino-acid-rich secretions (p.73).

Amongst the moths, feeding wholly within plant stems or even tree trunks is a widespread habit, and even this is represented amongst the butterflies. The American giant skippers, genus *Megathymus*, feed inside the stems and roots of various plants, including cacti of the genus *Agave*. This is the plant used to make tequila and mescal, and traditionally these strong drinks are bottled with an agave skipper caterpillar inside. Why this is done is uncertain, but one rather fanciful suggestion is that by doing so the spirit of the plant is infused within the drink! But feeding on special plant parts or within them is by no means the most unusual source of caterpillar food.

The greatest departure occurs in the lycaenids (blues) and riodinids (metalmarks) that feed on ant brood or plant bugs. Two species of tropical lycaenids that are entirely predatory as larvae have already been described in the section on egg laying (see p.38–39). One of the best-known examples amongst temperate butterflies is the genus *Maculinea* (also known as *Phengaris*), a group of about half a dozen species of so-called 'large blues'. They exhibit a range of strategies. These include the extraordinary process of being 'adopted' by a worker ant and carried back into the nest. Once inside, depending on the species, the butterfly larva either feeds directly on ant grubs, or persuades the ants to feed it, as if it were a member of the colony. Such deceptions depend on sophisticated chemical communication, and each of these butterfly species can normally only develop within the nests of one particular type of ant, to which it is specifically adapted and on which it is wholly dependent. One of the most striking things about such butterflies is the extent to which their growth is concentrated into the final instar. Some species of large blue put on 98% of their body weight during this stage.

As yet there are many species of butterflies about which we know little or nothing of their early stages or larval feeding habits, so there are sure to be more unexpected findings to come. For example, in North America the two species belonging to the genus *Neophasia* feed on pine tree needles, and are the only pierid butterflies known to do so. Were this not already established, such a discovery would be a great surprise.

BELOW The large blue, *Maculinea arion*, starts by feeding on thyme flowers. But the final instar must be adopted by a *Myrmica* ant, and taken to its nest. Once inside, the pinkish caterpillar starts to feed on ant grubs and pupae.

FROM JAWS TO STRAWS

The most obvious transformation that occurs when a caterpillar becomes an adult butterfly is the gain of wings and thus winged flight. However, equally striking from a functional viewpoint is the switch of feeding habit, from chewing solids to sucking liquids. During metamorphosis there is a massive reorganisation of the gut to deal with this change of diet and a total rebuild of the feeding apparatus.

In effect, adult butterflies feed through a drinking straw. At first sight the proboscis seems to be a simple tube, but appearances can be deceptive. The organ is made up of two separate tubes, each with a nerve and air vessel running within, held together by tiny hooks. But the food is not drawn through these tubes. Instead, each has a deep groove running along its entire inner surface, so that when the two are zipped together by the tiny hooks, a third tube, the actual feeding channel, is formed between them. This arrangement has the great merit that the two halves can be separated for cleaning – essential if the butterfly has been feeding on something sticky or lumpy. A bag inside the butterfly's head, with muscles that enable it to expand and contract, provides the sucking action. When not in use the proboscis is neatly coiled up, like an old-fashioned watch-spring. Microscopic examination of the 'tongues' of a wide range of butterflies reveals four subtly different designs corresponding to four major classes of feeding type: nectar, rotting fruit, hard fruit and pollen feeding.

BELOW This picture of a small tortoiseshell head was taken using a scanning electron microscope. Between the huge compound eyes the rolled-up proboscis is clearly visible.

LEFT A thirsty female of the mocker swallowtail, *Papilio dardanus*, with her tongue literally hanging out.

RIGHT Exquisite iridescent colours glitter from the mostly bare wing membranes of this glasswing, *Greta oto*, a close relative of the milkweed butterflies. When sucking nectar actively, the proboscis of a butterfly is invariably curved through about 120°, as seen here.

A butterfly on a flower is a familiar sight. Both sexes of many species feed avidly at a wide range of nectar-producing plants, including numerous daisies, verbenas, mustards and legumes. At this point the reader may ask, if it is true that adult butterflies do not grow, why do they feed at all? There are at least six reasons: water, energy, reproduction, defence, communication and neuro-transmission. Nectar can be very helpful with the first two needs. Insects are small and therefore have a relatively large surface area over which to lose water, and are thus in constant danger of drying out. Butterflies do drink plain water, but nectar can often provides a sufficient water supply. However, nectar is primarily a wonderful source of sugars, needed to power active flight, or for conversion to fat reserves prior to hibernation. Butterflies that migrate long distances, such as the monarch, depend on nectar as a source of energy.

BEYOND NECTAR

Rotting fruits can also supply sugars and alcohols for energy, but in addition they are a good source of proteins that can help with reproduction. As already discussed (p.27), males produce spermatophores that may be as much as 10%, or even more, of their body weight. These not only contain active sperm, but also nutrients that are passed to the female so that she can produce a greater number of mature eggs. She assists in this process through her own active feeding. The majority of butterflies that regularly feed from flowers rarely visit fallen fruits (the red admiral, *Vanessa atalanta*, is an exception). Likewise, most of the significant minority that do specialize on rotten fruit rarely visit flowers. Surprisingly, as yet we know very little of the ecological consequences of these preferences. Variations on rotten-fruit feeding include attraction to carrion, mammal dung, bird droppings and fungi.

Milkweed butterflies are strongly attracted to sources of pyrrolizidine alkaloids (PAs), the poisonous chemicals that make ragwort so dangerous to horses. Plants that contain these chemicals, such as certain borages, daisies (including ragwort) and legumes, may release them in nectar, but they are more concentrated in the leaves and roots. In the Philippines I observed *Ideopsis juventa* repeatedly probing the roots of *Ageratum*, a PA-containing daisy that had been upturned by a plough. Michael Boppré has filmed an even more direct assault on such plants, showing how African *Tirumala petiverana* use their sharp tarsal claws to scrape holes in the leaves to get at the PAs within. To take up the chemicals, the butterflies regurgitate a little fluid from the tip of the proboscis, mix it around until the PAs are partly dissolved, then suck the fluid back up again.

Remarkably, in the case of these butterflies, the special chemicals are used both for defence and to make sex pheromones (p.70). Milkweed butterflies are often attracted to dead individuals of their own group, probing them for PAs, and this specific chemical attraction has been confirmed by use of purified PA-extracts.

OTHER SPECIAL NEEDS

Growth and nerve function in insects are affected by sodium. Most land plants can be poisoned by relatively small amounts of salt, and so a solely plant-based diet will generally be very low in sodium. Recent research by ecologist Emilie Snell-Rood and colleagues has shown marked effects of common salt added to butterfly host-plant leaves – including greater brain development in female butterflies and increased thoracic muscle mass in males.

The special need for sodium may help explain why any concentration of common salt is very attractive to the adults of many butterfly species, especially males. They will sometimes flock in thousands on damp sand or mud, especially if enriched with urine, to form so-called 'mud-puddle clubs', where they may suck the tainted water for hours on end. It has been shown, in some cases at least, that sodium obtained in this way can be passed on to the females through the spermatophores. Butterflies in the tropics or in butterfly houses are often attracted to sweaty clothes and our skin, another potential source of sodium.

However, there is still much that we do not know about mud-puddling. Particularly mysterious is the phenomenon of fluid 'recycling'. It is often observed that as fast as the muddy water is sucked up by the proboscis, jets of fluid are ejected from the anus. In some species the proboscis is repeatedly extended backwards between the legs and some of the ejected fluid sucked up, and thus 'recycled'. Almost certainly there is more to mud-puddling than just the acquisition of sodium, and perhaps this is a way of concentrating some scarce but vital nutrients.

Species related to the holly blue are strongly attracted to damp, fresh wood ash. Wood ash is, of course, a good source of potassium – another chemical element, like sodium, that is essential for neurotransmisson. Whether or not the attractiveness of various other substrates has a common cause is unknown. Certain butterflies are attracted to damp concrete, which they will probe repeatedly – perhaps for calcium, also vital for nerve function. Other sources of attraction include aphid honeydew and plant sap. The European purple hairstreak, *Quercusia quercus*, seems to be addicted to honeydew, and the adults of an American relative, the

BELOW Full house! A mud-puddle club with males of three species of sulphur butterflies (Pieridae) literally packed cheek by jowl, as they probe the damp soil for salt and any nitrogen they can get for good measure.

Colorado hairstreak, *Hypaurotis crysalus*, may feed exclusively on sap oozing from the twigs of their larval host, the oak tree *Quercus gambelii.* These cases almost certainly represent alternatives to nectar as energy sources.

Arguably the most striking special need occurs in the passion vine butterflies. Members of a specialized subgroup of *Heliconius* have managed to use the proboscis, so beautifully adapted for feeding on liquids, as a clumsy but effective means of consuming pollen. The butterflies specialize on particular flowers, which they visit daily. They gather a ball of pollen and hold it within the lightly coiled proboscis. Onto this they regurgitate saliva containing a digestive enzyme. After some time the fluid, which by now has free amino acids dissolved in it, is sucked back and the spent pollen ball discarded. Adult *Heliconius* can live for many weeks or even months, females continuing to mature and lay just one or a few eggs per day, if they have sufficient resources. To do this they are very dependent on pollen as a source of amino acids for making their egg proteins.

CHAPTER 5
Flying

STATE-OF-THE-ART FIGHTER PLANES ARE 'fly by wire': they depend on continuous adjustment of the wing surfaces by advanced computer control. Without this technological support, such aeroplanes fly like stones. Insects, however, manage the complex manoeuvres of flight instinctively, using 'fly-by-wire' aviation skills that evolved well over 100 million years ago.

FLYING EQUIPMENT

The wings of a butterfly, as in other insects, are outgrowths of the second and third sections of the thorax, the major part of the body immediately behind the head. Behind the thorax comes the abdomen, which is primarily concerned with digestion, excretion and reproduction. Other than providing a connecting link between the head and the abdomen for nerves, blood and guts, the primary role of the thorax is locomotion. In addition to the wings, each of the three thoracic sections bears a pair of legs. The thorax is packed with the muscles that power active flight.

The size and shape of a butterfly's wings largely determine the rate at which the wings beat and can affect speed and manoeuvrability. During flight the wings flap, bend and twist in the complex ways necessary to achieve lift and movement as well as control. Robin Wootton, a palaeontologist specialising in the biomechanics and evolution of insect flight, has suggested that it is better to compare a butterfly wing with a sail, rather than the fixed wing of an aeroplane or the muscular wing of a bird.

The wing veins form a crucial part of the system, as do the little spars and levers at the base of the wing

OPPOSITE A poplar admiral, *Limenitis populi*, at the moment of take-off. After a rapid upstroke, the broad wings are rotated sharply downwards to give maximum lift, almost literally flinging the insect into the air.

BELOW Doing the locomotion: a pair of legs arises from each of the three segments of the thorax, and from the middle and hind segments the all-important wings (left legs not shown).

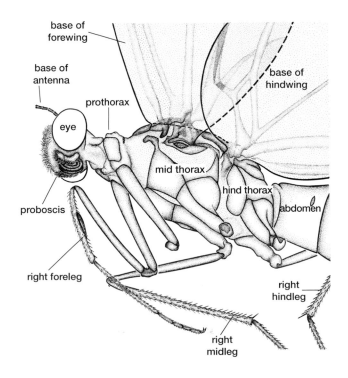

base of forewing
base of antenna
prothorax
base of hindwing
eye
mid thorax
hind thorax
proboscis
abdomen
right foreleg
right hindleg
right midleg

through which control is achieved. Two very large sets of opposing muscles bring about the up and down strokes by pulling against the elastic walls of the thorax, deforming them, and thereby transmitting power to the wings. Regulating how all these systems interact is arguably the greatest achievement of the insect brain.

PERFORMANCE

What sort of flight performance does a butterfly achieve? Houseflies are perhaps the insect masters of the air, being able to fly backwards and even upside down in a fully controlled way. Butterflies are not quite that good, but they are nonetheless remarkably manoeuvrable. In the deserts of Palestine, biologist P.A. Buxton observed a little blue butterfly, *Chilades galba*, flying within the bounds of a rest-harrow plant during a desert storm. The plant was no more than a foot across, but the *Chilades* was able to fly freely within its confines, while all other insects had long ceased to be active. Fast flyers, such as *Charaxes* or *Graphium*, experience acceleration forces equivalent to several times the pull of gravity when they make lightning-quick changes in direction to avoid attack. Some butterflies can hover in one spot for minutes on end, while others sustain forward flight for hours, travelling 100 km (60 miles) in a day. Sprint speeds of 30 km per hour (20 miles per hour) or more are reached by certain butterflies, which places them, along with dragonflies and hawkmoths, amongst the fastest of all flying insects.

BELOW **These male Rajah Brooke's birdwing,** *Trogonoptera brookiana*, **found at Kuala Wah, Malaysia, have been disturbed while drinking from wet sand. As they fly up, caught by the photographer's flash, drops of water fall from their bodies.**

RAPID FLIGHT

Butterflies and their caterpillars face the continual threat of becoming a predator's next meal and they employ a wide range of strategies to avoid this. Adult butterflies may try to escape from their enemies through sheer speed. Perhaps the most charismatic of all African butterflies belong to the genus *Charaxes*, a group of over 150 large, powerful and very beautiful species. In 1972, I was in Angola, seeing tropical Africa for the first time, and discovering what makes *Charaxes* so appealing to the field naturalist. Even where they are common they are almost impossible to see. You have to become a hunter to engage with *Charaxes* and learn to exploit their Achilles' heel: a fatal weakness for rotting fruit, dung or carrion.

LEFT **Among the fastest of butterflies, this *Charaxes jasius* is feeding from rotten banana, apparently oblivious to the ants sharing his feast. The males are hill-toppers that perch, and fly out at great speed in pursuit of rivals.**

RIGHT Crow butterflies, *Euploea*, in flight, revealing different positions of the wings as they complete the upward-and-downward flapping cycle.

Adult *Charaxes* are attracted to these apparently disgusting sources of nourishment like wasps to a picnic. Collectors make traps baited with fermenting bananas and within hours can lure hundreds of these superb butterflies. On one occasion, as I approached an open spot amongst the undergrowth where I could see some swallowtails feeding (like the *Charaxes* they have a liking for dung and carrion), I glimpsed a whitish-green butterfly – but then suddenly it was gone. What I later discovered to be *Charaxes eupale* shot up into the trees like a bullet, flying in a dead straight line at about 60° to the ground. Estimating the speed of insects is notoriously difficult, but my guess was 50 km per hour (30 miles per hour). It reached the safety of the forest canopy high above within a couple of seconds. Like most creatures, butterflies are vulnerable to attack when they are feeding. The all-round vision provided by their big compound eyes, which are extraordinarily sensitive to movement, gives early warning. At the slightest movement they are off, catapulting forward with fantastic acceleration. After a frantic dash to a sheltered spot, a butterfly will settle within an instant, leaving its would-be attacker, even a bird capable of greater speed, bemused as to where it went. But fleeing from danger in this way is not a strategy adopted by all.

STYLE

Individual butterflies can adopt different styles of flight – fast, slow, gliding, twisting or hovering – for different purposes. Even so, different groups and species can often be recognised by their general flight style. Differences in wing and body shape have a major influence, as in the fluttering, damsel-fly-like progression of the dragontails, genus *Lamproptera*, with their broad hindwing tails that can be longer than the rest of the body. Style can also be influenced by more ancient inheritance.

LEFT A green dragontail, *Lamproptera meges*, with its remarkably long hindwing tails, is reflected on a watery surface as it probes in some mud. Photographed near the famous waterfall at Bantimurung, southern Sulawesi, Indonesia.

The erratic, tumbling flight of white butterflies is characteristic of many species belonging to the group, as is the fluttering half-hover, half-settled movement of most swallowtails when taking nectar. Many skippers beat their wings very fast, making a whirring sound as they pass. Admirals and many other butterflies of the Nymphalidae family have an elegant flap-and-glide, swooping style, whereas browns, in the subfamily Satyrinae, often adopt a bobbing flight. These differences in flight patterns appear to be reflected by differences in the anatomy of the wing bases, but the importance of the anatomical variations has not been proven.

FINDING RESOURCES

The greatest benefit of flight is the ability it confers in finding resources. As biologist Larry Gilbert discovered in Parque Nacional Corcovado, Costa Rica, if you follow a colony of long-lived *Heliconius melpomene* butterflies for some weeks, you will find that they have a daily rhythm. The females visit scattered passion vines in a sequence, to check their suitability for egg laying, and they also visit particular *Psiguria* flowers daily, to feed on pollen and nectar. Up to 80% of their total egg production is dependent on the nutrients obtained from pollen. Males follow a similar routine, visiting the host plants in the hope of finding virgin females and the same flowers for food. As fresh butterflies emerge, some will join in the daily flights, learning the locations of these vital resources from their parents, grandparents and even great grandparents. This collective 'traplining' is about as social as butterflies get, but it also underlines the importance of flight as a means of finding three vital adult resources: host plants, mates and food. However, flight can have other vital functions, too.

LEAVING HOME

Amongst the seasonal butterflies of temperate regions, the smaller males usually complete development a few days earlier than their females. Rather than just 'hanging around', they fly off in search of other places where their own species may be found. The females they leave behind, which will emerge a few days later, will usually be courted and mated by males from elsewhere. In this way inbreeding is largely avoided.

However, it would be wrong to suggest that all females are 'stop at homes'. There are good reasons why they may find it necessary to move on, too. Some butterflies feed on plants that flourish in a particular habitat for a limited amount of time. For example, in northern European woodlands the heath fritillary, *Mellicta athalia*, is dependent on cow-wheat, a semi-parasite of grasses. As the woodland matures the grasses and cow-wheat become shaded out and the average temperature at ground level falls because of the lack of sunshine. The butterflies must then move on to find a new, more open patch of woodland, where there is cow-wheat for their caterpillars to feed on.

Curiously, for a species that is regularly faced with this problem, the heath fritillary seems to have very limited powers of flight. While a particular wood remains suitable, most of the butterflies live and die virtually within sight of where they were born. But as the shade increases and the females sense that it is time to move on, they manage to fly only a couple of kilometres at most. Within that maximum range they must find a suitable place to colonize, or die without being able to contribute to the next generation. Some butterflies, however, are faced with even more immediate problems of life and death, caused not by gradual changes to their home territory, but by the inevitable onset of winter. Once again flight comes to the rescue.

ABOVE **A pair of heath fritillaries,** *Mellicta athalia.* **These butterflies live, barring accidents, for about 5–10 days. During that time most individuals will not move more than about 150 m (500 ft) from where they emerged from the pupa.**

THE VALLEY OF PURPLE BUTTERFLIES

In the winter of 1971 a totally unexpected discovery was made on Taiwan. The central mountain range runs almost the whole length of the island, with over 130 peaks rising above 3,000 m (9,800 ft). At the southern end of the range the mountains are lower, with sheltered valleys running down to the sea in the east and lowlands around Chow-Jou to the west. Walking into one of these densely forested south-western valleys, a Taiwanese entomologist came across a wonderful

ERRATIC FLIGHT

ABOVE When it comes to feeding or courting, the small white can fly as well as any other butterfly. Here a male hovers behind an unreceptive female.

The small white, *Pieris rapae*, is very conspicuous as it crosses fields and gardens. Its flight seems rambling, haphazard and uncontrolled. You might think it was drunk, as it flaps along in a seeming daze, first this way then that. But this erratic, rock-'n'-roll style offers at least two advantages. The unpredictable zig-zag flight makes it very difficult to catch. A small bird in pursuit must anticipate the butterfly's trajectory. It is necessary to aim a precise distance ahead of where it is seen at any given moment, as during the time it takes to make the final lunge the butterfly will have moved on. A succession of sharp, rapid and seemingly random changes in direction make it almost impossible to estimate where the butterfly will be in the next instant. The second advantage relates to the butterfly's remarkable ability to find scattered foodplants: no cabbage is safe.

BELOW Four species of crow butterfly hang in the trees, wings folded. Every winter successive generations return to the same Taiwanese valleys, now known as the 'valleys of the purple butterflies'.

sight: the trees all around were aglow with a violet sheen. They were covered in tens of thousands of blackish-purple 'crows' – four species of milkweed butterflies belonging to the genus *Euploea*. Since then, three more butterfly valleys have been found on the island, all in the south. Every November butterflies appear at these same locations. What are they doing there and where do they come from?

Although the whole island of Taiwan extends only 394 km (245 miles) from north to south, the onset of winter produces a very marked north–south temperature difference. February temperatures can be as much as 4 °C (7 °F) lower in the north. Like all milkweed butterflies, *Euploea* are not frost-hardy: air temperatures below +4 °C (39 °F) can be lethal, especially if the butterflies are wet. In late autumn, the purple crow butterflies in the central and northern parts of the central mountain range start to fly south. At first they make little bands, flying down into the nearest valleys. As cold air starts to arrive, flowing out from the Chinese mainland, the butterflies hurry on to lower and warmer glens. As they descend they form larger and larger bands. As entomologist Hsiau Yeu Wang described it, the process is like a snowball rolling downhill, and spectacular numbers of butterflies can sometimes be seen flying past. By November, huge groups begin to stream into the lowest and warmest valleys at the southern tip of the island, settling in the trees at 200–350 m (650–1,150 ft) elevation.

The aggregating butterflies spend most of their time resting on the leaves of the trees. Sometimes they are so densely packed that the trees are covered, the butterflies literally standing wing to wing on the twigs and branches. On warmer days they become active, visiting nearby streams during the morning to drink and returning to their roosts by midday. When a cold front arrives, they will sit motionless for days on end. But if the air temperature drops well below 4 °C (39 °F), disaster strikes. The cold butterflies lose their grip and, one by one, fall to the forest floor, which is soon carpeted by their dead bodies. However, in most years survival is good. As the air starts to warm up again in late March, the butterflies become more active and search for nectar to replenish their energy stores. Courtship follows, and mated females gradually begin their northerly return flight, to lay their eggs in the higher valleys of the central and northern parts of the island. Five or six months later, before the return of winter, their grandchildren or even great grandchildren will start the southerly migration again, completing the cycle and filling the valleys with purple butterflies once more.

THE MIGRATING MONARCH

The purple butterfly valleys of southern Taiwan are beautiful to behold, but for sheer scale they pale to insignificance compared with the migrating bands of North American monarchs, *Danaus plexippus*. Every autumn, all the monarch butterflies of North America, east of the Rocky Mountains, fly southwards. At the latitudes between Boston Massachusetts and Birmingham Alabama, not a single

LEFT Monarchs circle above the trees on a warm day near Michoacan, their Mexican overwintering site. Tens of thousands take to the air on such days, to find water, which they need if they are to survive the winter months.

monarch butterfly, pupa, caterpillar or egg remains to pass the winter. In most years the number of monarchs heading south is greater than the entire human population of the USA – but where are they going? The goal of each butterfly is to find one of about 30 particular points in the mountains 60–160 km (35–100 miles) west of Mexico City. These little forest patches at 2,400–3,500 m (7,850–11,500 ft) elevation, amounting to no more than 25 hectares (60 acres) in total, offer a safe haven to pass the winter months.

Like their relatives in Taiwan, the monarchs cannot endure freezing, and they must make this annual pilgrimage – a flight of up to 5,200 km (3,200 miles) – to survive. Although the mountains are cold, with frosts and even occasional snow, the temperature rarely drops below 5 °C (41 °F) at the overwintering sites. At just one of these special places, densely clothed in oyamel firs, more than 10 million monarchs may assemble. They hang silently on the trunks and branches of the trees, which may bow under their massed weight. On warmer winter mornings, when many of the roosting butterflies venture out to seek water from nearby streams, the sight of tens of thousands of monarchs circling the air is breathtaking.

Exactly how the monarchs manage to find a few tiny spots on the vast map that is North America still defies our full understanding. Whatever the navigation mechanism, the whole phenomenon raises numerous questions about their flight. How long do the butterflies take to make the journey? Do they migrate by night, by day or both? Do they cross the sea, or fly overland? How much energy do they need and do they refuel on the way? Why don't they migrate to somewhere warmer? When do they mate? When do they lay eggs? What happens on the return flight? We do have answers to some of these questions, thanks to pioneers like zoologist Fred Urquhart and the many painstaking studies of Lincoln Brower and other entomologists.

THE LONGEST OF JOURNEYS

If a monarch could fly at 20 km per hour (12 miles per hour), 12 hours a day, seven days a week, it could travel 5,000 km (over 3,000 miles) in just three weeks. However, it seems that the journey from north of the Great Lakes to central Mexico takes longer than that. Butterflies that mature in August start the great trek south as early as the beginning of September. Before doing so, they feed avidly to build up fat reserves and perhaps flight muscle, increasing their total body weight by as much as 30%. As they fly south they roost each night, a few dozen or a few hundred sleeping together, often in favoured trees or groves that are used year after year. They also have to stop to take on more nectar, and they make use of favourable winds by soaring high into the sky. Monarchs can cross huge expanses of sea, but the normal migration route for most of them passes south-west, to funnel through Texas. Finally, the butterflies arrive at the overwintering sites during November. So on average they probably travel about 80 km (50 miles) per day.

CHOICE OF SITE

After their momentous journey, why do the monarchs choose to overwinter in a place so high in the mountains, where the risk of freezing is still considerable? The answer lies in the fact that the butterflies have to be able to survive until their North American territories become warm again. They do this by entering a state of suspended animation, which requires a site that remains cool but is unlikely to freeze. In warm or hot lowlands, the butterflies would burn up their fat reserves long before the return of spring, and they would be overcome either by starvation or old age.

MATING MONARCHS

Throughout the journey south the butterflies rarely engage in mating; sex is something to be saved for the spring. By the beginning of February temperatures are rising, and daylength has increased to more than 11 hours, the critical period that triggers onset of renewed migratory activity – and sexual pursuits. Mating among these sex-starved insects at the overwintering sites appears brusque. A male grasps a female in flight, overpowering her with his long, spiny legs. Wings outstretched, he stops active flight, and the pair glide to the ground. Once they have landed some matings do occur, but the females often successfully resist. This is because, it seems, such desperate males are in poor condition, and unlikely to be fit enough for the journey north.

By early March the colonies finally break up, and the butterflies stream north on the return migration. Within a few days most of the spent males die, but the females and the fit males carry on, flying to the middle states of the USA, where the survivors mate and the females then lay their eggs on suitable milkweed plants wherever they can find them. By May the monarch will have repopulated the whole of its breeding range, right into Canada, some making the whole return journey, but mostly by new broods moving ever northwards before reproducing. In fact very few of the overwintering females live long enough to reach so far north; butterflies seen in the most northerly regions are likely to be their sons and daughters, or even grandchildren. When September returns, it is the third or fourth annual generation that makes the next great trek south.

BELOW Mating monarchs at Pacific Grove, February 1996. The male is to the left. Monarchs west of the Rockies overwinter on this part of the California coast, instead of flying to Mexico, as famously but somewhat inaccurately recalled in John Steinbeck's novel *Sweet Thursday.*

WHY DO MONARCHS MIGRATE AND HOW DO THEY NAVIGATE?

Why monarchs migrate in North America may not only be a question of frosts. Exactly how the butterflies find their way across a whole continent is not fully understood either. All that seems certain is that the butterflies instinctively know the correct direction, and find their way using time-compensated information gathered from the sun's position, and sensitivity to the Earth's magnetic field. Monarchs use their magnetic sense to maintain direction even on overcast days.

One surprising idea is that the 'brusque' manner of monarch mating behaviour, observed at overwintering colonies, is linked to long-distance migration in a way beyond just the desperation of 'spent' males. Most milkweed butterflies use complex pheromone systems and elaborate behaviour to persuade their females to mate (p.70). Male monarchs at overwintering sites seem to dispense with all such sophistications, and literally grab females in flight (and often males by mistake!) and then try to overcome rejection by sheer physical force. Both sexes have striking anatomical modifications that make such 'rough sex' possible – including the long and spiny male legs already noted. However, not only the migratory monarchs of North America, but also the non-migratory monarchs of Central America and the Amazonian region have exactly the same anatomy. Moreover, the monarch's two closest relatives, a smaller Caribbean species and the southern monarch of Chile and Argentina, have exactly the same modifications. The West Indian species is non-migratory, but the southern monarch, *Danaus erippus*, is believed, although poorly documented, to undergo annual migrations. No other milkweed butterflies have any of the 'rough sex' anatomical specialisations of these three. Maybe the ancestors of this entire little group were migratory – and this somehow affected their mating system.

Another recent suggestion is that monarch migration is connected with 'escape' from the parasitic disease 'Oe' (p. 19). Low numbers of the parasites cause little harm, but heavily infected larvae give rise to smaller adults with reduced flying ability. Even greater infections cause deformity and death. So migration could be a purging system; only the best and most vigorous monarchs make it all the way to Mexico, and they would carry few parasites, thus 'cleansing' the population once a year. Nothing is yet known about 'Oe' in areas where monarchs do not migrate, or where the southern monarchs occur. And what appears to be the same parasite has been found in other *Danaus* species.

Evolutionary speculations aside, even the 'technical' means whereby the remarkable feat of migrating thousands of kilometres to a tiny area of Mexico is achieved are not fully resolved. Although a very active area of research, much remains to be done if we are to understand exactly how butterflies that have never been to their destination before – and will never go there again – are so successful in finding their way. The spectacular flight of the North American monarch, now greatly threatened by genetically modified crops, deforestation and climate change, is one of the most remarkable chapters in the biology of butterflies.

CHAPTER 6
Communicating

LIKE HUMANS, BUTTERFLIES HAVE AT LEAST five senses: taste, touch, hearing, smell and vision. As we have seen, in addition, the monarch appears to be sensitive to and make use of information from the Earth's magnetic field (see p.65), and other species may have this ability too. All five primary senses can be used in communication, but vision and smell seem to be the most important.

OPPOSITE A pair of courting dingy skippers, *Erynnis tages*, photographed in the French Pyrenees. The female is above, the male below. Courtship is, with rare exceptions (e.g. p. 29), all about communicating (pp. 23–26). All senses can be involved, but scent and sight seem the main communication 'channels' in butterfly courtship. Butterflies also need to communicate with other organisms.

LEFT A cracker, *Hamadryas februa*, perches head-down on a favourite tree, ready to dash out with a noisy welcome for any intruder – or perhaps a potential mate. Members of the genus *Hamadryas* probably qualify as the world's noisiest butterflies.

TASTE

Butterflies can taste with their feet. To see this you only have to hold a hungry butterfly carefully between forefinger and thumb, and then place it, feet down, against a tissue moistened with sugar solution. Almost without hesitation the butterfly will uncoil its proboscis and start to feed. Butterfly feet are equipped with sugar receptors. The feet of many females also appear to have specialised taste receptors to detect chemicals that characterise their host plants. Before laying they will 'drum' a leaf (see p.34) and only if their scratching releases the correct compounds will they continue.

TOUCH

A sense of touch is probably important not only for day-to-day survival, such as take-off and landing, but also in courtship. The two prominent antennae that arise just above the eyes are often waved about as a butterfly investigates something (they are often referred to by children as 'feelers'). Their main purpose is smell (and in the monarch they are the site of a time-compensating 'clock' used in orientation), but the antennae are also sensitive to touch. During the courtship of some species, the male will stroke or tap the female with his antennae, and this may indicate communication by touch as well as smell.

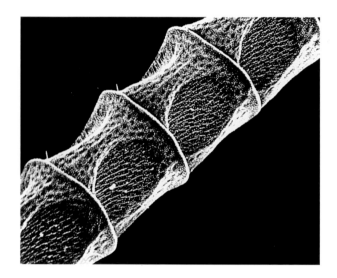

ABOVE Four segments of the antenna of a *Calinaga* butterfly (Nymphalidae) from China, as seen under the scanning electron microscope. The large pits are filled with sensors that detect scents, the antennae being a butterfly's main organs of smell.

HEARING

That butterflies can hear will come as no surprise to anyone who has tried to photograph a tortoiseshell on a flower, using an old-fashioned camera held close. The sound of the shutter is sufficient to alarm the butterfly, its almost instant departure often leading to a picture with only a 'smudge', or even just the unadorned flower. Given an apparently good sense of hearing, it is perhaps surprising that few adult butterflies appear to use sound for communication.

One potentially good example is offered by the South American 'crackers', genus *Hamadryas*, familiar to Charles Darwin. These butterflies rest head-down on tree trunks and if approached fly out, making an audible clicking sound. Clicks are also emitted if two *Hamadryas* encounter each other, but the real significance of the display is still unknown – possibly both courtship and territoriality are involved. Even the mechanism by which the sound is produced is still not reliably understood. The butterflies have a special 'ear' located in the forewing, which may be the primary means of hearing their own clicks, and those of other males.

SMELL

Scent is very important to adult butterflies. They can find sources of food such as rotting fruit, nectar, various chemicals such as alkaloids, and probably even water, by smell. The main receiving organs are the antennae, which have many tiny receptors connected by nerve fibres to the butterfly's brain. Electrical recordings have shown how particular receptors are usually sensitive to just one chemical, or a few related groups of chemicals. For example, many butterflies have receptors sensitive to vanillin, a substance commonly found in nectar.

Male moths are famous for being able to find their mates in the dark, in response to specific volatile chemicals or mixtures of chemicals. The females release these pheromones into the air when they want to attract a mate. Although male butterflies nearly always locate their mates visually, the females do emit odours, and in some cases these are important for bringing the sexes together, as in those species of *Heliconius* that engage in 'pupal rape' (see p.29). Male *Heliconius* use the female odour to find pupae that are about to emerge, without being able to see them from a distance. The same also appears to be the case in the beautiful Australian lycaenid *Jalmenus evagoras*, studied in detail by biologist Naomi Pierce. Many males can be attracted to a single female pupa and wrestle among themselves in apparent hope of being the winner.

The use of scents by males appears to be of major importance in butterfly communication. A remarkable variety of so-called androconial organs only occur in males and, although most of these have never been investigated properly, it is widely assumed that they are used to store and release scents. Whenever they have been studied in detail, this has proved to be the case. Some examples have already been mentioned (see pp.23, 25), regarding a role in courtship, such as the special fluffy hindwing scales of male *Troides*.

ABOVE **A lust for fruit and a keen sense of smell can get butterflies into trouble. This Rydon trap was set by the author in Angola in 1972. Baited with fermenting banana, it has ensnared numerous** *Charaxes, Neptis* **and other fruit-feeding nymphalids.**

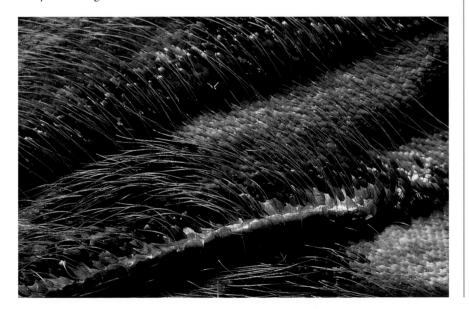

LEFT **While some scent scales are hidden inside pockets, or consist of brushes to be fanned out when needed, many occur on the open wing surface. Here is a detail of the androconial scales from the forewing of a male fritillary.**

VISION

Vision, as might be imagined for such colourful creatures, plays a major role in the life of adult butterflies. Butterflies can see a little further into the red end of the spectrum than humans and well into the ultraviolet. Recent work suggests that colour sensitivity varies across different groups, and eventually this may offer some explanation for the behavioural differences observed between species.

An appreciation of colour affects butterflies not only in the search for nectar and in mate location and selection, but also in male–male interactions. Collectors have long exploited the fact that the males of many species are readily attracted to almost anything vaguely butterfly-like that matches the butterfly's own principal

THE PLAIN TIGER'S CRUCIAL SCENT

ABOVE All male milkweeds have abdominal pencils similar to those of the plain tiger. These fully everted pencils belong to a species of the American genus *Lycorea*.

RIGHT The hair-pencils of *Euploea* are brilliant yellow. Unlike the plain tiger, neither *Euploea* nor *Lycorea* use transfer particles, but they do make pheromones from PAs.

The most detailed studies on butterfly scent organs have been made on various species of the milkweed group (Danainae). The male plain tiger, *Danaus chrysippus*, has two scent organs: a brush of special scales held within a sheath close to the tip of the abdomen, and a pocket-like wing organ that opens near the middle of the upperside of the hindwing. As in most danaines, male plain tigers are strongly attracted to sources of pyrrolizidine alkaloids (PAs), such as borages and various other plants (see p.26). Before courtship, a male plain tiger will thrust out its abdominal pencils and push them into the hindwing pocket, where the pencil hairs come into contact with a glandular area covered with androconial scales. This 'contact behaviour' stimulates the production of the pheromone danaidone, which is made from the PAs by means of an enzyme

within the wing pocket. Danaidone, together with other chemicals, is transferred to the female antennae during courtship by means of tiny pheromone-transfer-particles produced in the male's abdominal brushes. Males denied access to pyrrolizidine alkaloids, or prevented from making contact behaviour, make little or no danaidone and are very rarely successful in courtship.

As danaidone or a similar substance is produced by all danaines, what message does this pheromone convey? The concentration of danaidone can signal the male's 'fitness' and quality of his spermatophore. The spermatophore brings to the female more than just sperm — it supplies quantities of the alkaloid that help protect the female herself, as well as her precious fertilized eggs: PAs repel many spiders, bugs and other deadly egg predators (see pp.28, 49).

colour. If you want to catch a male of a butterfly that is blue, try pinning a piece of shiny blue paper to your hat. This works well for attracting patrolling species, such as *Papilio ulysses* or blue *Morpho*. Even better, you can use a dead specimen with its wings partly open.

As already recounted, 'mud-puddling' is an activity that many butterflies, especially males, engage in. If you come across a large mud-puddle-club with many individuals of several species, and the group has not been disturbed, you are likely to find that the differently coloured individuals and species are segregated: all the white ones together, the orange ones together, blue ones together and so on. The simplest explanation for this relates to the idea of flocking or herding, and safety in numbers. Certainly being faced with an array of similarly coloured creatures seems to make it more difficult for predators to get a 'fix' on one particular individual and carry through a successful pursuit. Thus male butterflies may use their bright colours to communicate among themselves. But what of the obvious alternative: that the bright colours of male butterflies are used to impress their potential mates?

COLOURFUL ATTRACTIONS

The idea that male butterflies use their bright colours to attract a mate was favoured by Charles Darwin, but disputed by Alfred Russel Wallace, and was one of the few facets of evolutionary biology on which these two pioneers seemed to disagree. However, nearly 100 years passed before a good experiment was performed to investigate this aspect of butterfly biology. The experiment was preliminary, and the entomologist who devised it, Bob Silberglied, died tragically in an air crash, soon after. Little follow-up work has been carried out, even though his experiment was very revealing.

Anartia amathea is a common nymphalid found throughout much of Latin America, including Panama. The butterfly is largely black, with a red stripe across both wings. The sexes are essentially similar, although the males look a little brighter. Working in Panama, Bob took three equal-sized groups of fresh males and painted out the red stripes of one group with black ink. The second group he overpainted with a clear solvent used in the ink, and the third group he left untreated. He did exactly the same for an equal number of virgin females, and then released all of them into a large flight cage. He then simply observed any matings that took place.

The results were striking, even though the numbers were small (the experiment should be repeated many times, with various species). Quite simply, the males ignored all-black (non-red) females, but females were indifferent to the colour of their suitors. As you would expect by chance, the blackened males were involved in about a third of all of the matings observed. So in this case it looks as if Darwin was wrong. However, Bob Silberglied was also responsible for a major study

ABOVE A male *Anartia amathea*, photographed in Tobago, probably on the lookout for rival males or potential mates. Surprising at it may seem, Bob Silberglied's experiment suggested that the females will not be the least bit impressed by his gorgeous red coloration.

showing that, in some species at least, females *are* sensitive to male colour. This was the work on *Colias* (see p.24) that demonstrated that the ultraviolet coloration of males forms an essential signal if courtship is to be successful. From all this, and various other lines of evidence, we can tentatively conclude that both male and female butterflies can be sensitive to the colour of their prospective mates, but this is not always the case. Trying to draw grand generalizations in biology can be a frustrating business!

PARTNERSHIPS AND BETRAYALS:
ANTS AND BUTTERFLIES

Early in their evolution, the ancestors of two of the major groups of butterflies, the families Lycaenidae and Riodinidae, apparently developed a partnership with ants. If you watch a caterpillar of the North American Boisduval's blue, *Plebejus icarioides*, you will see ants gather round and begin to stroke it with their antennae. In response, the caterpillar secretes a sugary fluid that may be rich in amino acids. The ants eagerly collect this substance, released from special organs located on the caterpillar's back.

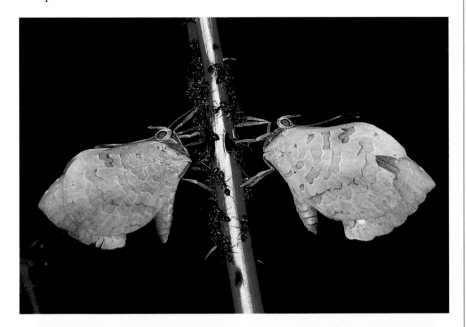

LEFT Two brownies, *Miletus ancon*, stand on their stilt-like legs, feeding on honeydew from plant bugs hidden under numerous attentive ants (*Dolichoderus*). Perfidiously, if they lay eggs here, their larvae will almost certainly eat the little providers.

LEFT *Dolichoderus thoracicus* crowd around a third instar larva of the butterfly *Miletus symethus* that is busily eating members of their aphid 'herd' (*Pseudoregma bambucicola*) – yet the ants do not attack it. Two hatched white eggs are also visible.

ABOVE A large blue *Maculinea arion* drinks nectar from wild thyme. Females lay their eggs singly, in a flowerbud of the same species. The larva must later be adopted by *Myrmica sabuleti* – no other species of ant will do.

This is not a one-sided deal. In return, the ants will usually protect the butterfly larvae and pupae from attack by wasps, bugs or ants – including themselves. In effect, the ants 'adopt' the caterpillars rather like the aphids and other plant bugs that they often corral and milk for honeydew. Any doubt that this partnership could really benefit the butterflies was removed by the work of Naomi Pierce, who demonstrated that without ant protection up to 95% of the pupae of an Australian lyacenid, *Jalmenus evagoras*, were attacked and killed by a parasitic wasp. With *Iridomyrmex* ants in attendance, the wasps were completely unsuccessful. Like the caterpillars, the pupae of *Jalmenus* also release amino acids that are highly attractive to the ants.

Species of blues that have such partnerships often have more than one sort of ant organ. In many cases these include a pair of little paintbrush-like structures that they can stick out at will, to emit substances that act like ant alarm pheromones. If the larvae find themselves under attack, the release of these mimic pheromones rapidly attracts any nearby ants to their assistance. The ants seem willing to do this, as they will be rewarded with the nutritious secretions.

Sounds are also produced by the larvae (and pupae) of many lycaenid and riodinid butterflies, apparently for much the same reason. The caterpillars of the metal-mark *Thisbe irenea* have two ridged rods, the vibratory papillae, located just behind the head. Tropical biologist Phil DeVries has shown that if the caterpillar moves its head back and forth, a rough area of the head rubs against the rods, to produce a low, grating noise. This sound is difficult for us to hear, but it is readily transmitted through the leaf or stem on which the caterpillar is standing, and is detected by *Ectatomma* ants, which produce similar sounds themselves. These sounds act in much the same way as the alarm pheromones, so that the ants are drawn to the caterpillar's defence.

THE RISKY BUSINESS OF BETRAYAL

Having successfully tapped into the communication channels of ants, perhaps it was inevitable that by exploiting these mimetic 'skills' some blues and metal-marks would find ways to betray the ants' trust. How some species eat bugs attended by ants, or even get into ant nests to eat ant brood or be fed by the worker ants, has already been described (see pp.39, 46). In the present context, it is notable that each species of *Maculinea* can only survive in the nest of the particular species to which it is adapted. Quite often *Maculinea* larvae are picked up by the wrong species of ant. Once in the wrong nest, the false pheromones that do not match those of the owners become the caterpillar's death warrant. Only when taken into the right sort of nest, where its subterfuge will work, does a *Maculinea* caterpillar have a good chance to thrive at the expense of its hosts.

CHAPTER 7
Varying

BUTTERFLIES HAVE ALL SORTS OF SPOTS and markings on their wings, which can often be used to identify and differentiate the various species. Sometimes the differences are so minute that they are almost invisible to the naked eye. At other times the differences are actually red herrings because they merely represent individual variation. In fact, butterflies constantly tax the brains of taxonomists and test their classifications because they vary so much: between individuals, between the sexes, with the seasons, and across regions, countries and continents.

OPPOSITE **Two-brand crow butterflies, *Euploea sylvester*, with one common Australian crow, *Euploea corinna* (second from bottom on right), and one swamp tiger, *Danaus affinis* (bottom left), gathered at an overnight roosting site, Queensland, Australia. The two-brands might all look the same, but no two individual butterflies, of whatever species, are ever identical in all respects. Individual variation is the rule, without exception.**

VARIATION ON A THEME

Throughout much of North America, Europe, China, Japan and even parts of Africa, you may encounter one of the world's most lively and attractive butterflies, the small copper, *Lycaena phlaeas*. Its forewings are mostly copper-coloured, but the hindwings are mainly blackish-brown, with just a short, marginal, coppery band. Sometimes you will come across individuals that have these markings, but with one striking addition: a little row of three or four bright blue spots on each hindwing.

Does this blue-spotted sort represent another species? In this case we know that it does not. For *L. phlaeas* these differences are just part of normal variation, rather like variations in eye and hair colour in our own species. What is the source of such variation, and how is it produced?

GENETIC DIVERSITY AND INDIVIDUAL VARIATION

All butterflies reproduce sexually, developing from the union of an egg and a sperm cell. Therefore each and every butterfly gets two sets of genetic instructions (DNA), one from the mother and one from the father. Like all complex organisms, the growth and development of an individual butterfly is strongly affected by literally thousands of these instructions, many of which can exist in two or more alternatives (such as 'make blue spots on hindwing' or 'don't make blue spots').

Through the process of gene recombination that occurs during the formation of egg and sperm cells, and subsequent fertilization, this genetic diversity gives scope for almost limitless individual variation.

GENETIC DIVERSITY AND SPECIES

One of the most distinctive butterflies in the world is the peacock, found in the UK and right across Europe eastward to Japan. Everywhere it looks the same, and seems instantly recognizable. Yet we can be sure that each individual peacock butterfly is genetically unique. So why don't they all look different? This is because genetic variation is only expressed within certain limits, in the same way that, despite the enormous genetic diversity that occurs in humans, we never have

RIGHT The peacock, *Aglais io,* with its fabulous eyespots and rich russet-red wing colour, is unmistakeable from Tottenham to Tokyo. Even the sexes are hard to tell apart without microscopic examination. But this does not mean that peacocks are genetically all the same.

RIGHT Striking variations of the peacock pattern do occur sometimes. This one is almost 'blind' on the hindwing, and the eyespots can fail on both wings. However, in this species such pattern variants must be at a strong disadvantage.

difficulty in recognizing any man or woman as a member of our own species. Natural selection can prevent variants from reproducing, either because they are poorly adapted, or because they have difficulty in attracting a mate. Such a selection process is called 'stabilizing', as it preserves the characteristic features of a species, such as the unique pattern of the peacock.

In fact, variations of the peacock do occur in nature, but they are rare. It seems that even if some survive and are able to find a mate, stabilizing selection rapidly stops their spread. In the peacock even the two sexes are similar: you have to be a bit of an expert to tell them apart (as an alternative to looking at their largely internal sex organs, you can do it by examining their front legs – always smaller in the males). However, this is by no means the case in all butterflies. Some species have different forms for males and females and others show great variability in the appearance of individuals.

TWO VARIETIES – ONE FOR EACH SEX

Many butterfly species show sexual dimorphism, with a distinct form ('morph') for each sex. In the case of the New Guinea paradise birdwing, the male is smaller than the female, exquisitely coloured in gold and green, and its hindwings have beautiful curved tails and a fringe of long, straw-coloured androconial hairs. In stark contrast the larger female is black and white with a little red, has evenly rounded hindwings, and no tails or striking fringes.

VARIETIES WITH OR WITHOUT SEXUAL DIMORPHISM

In various parts of Asia, such as some of the remaining forested hills in Sri Lanka, or overgrown gardens in Hong Kong, you may come across *Papilio* (*Chilasa*) *clytia*, a large butterfly belonging to the swallowtail family. This species is notable because it has two sharply distinct forms. One form is mostly blackish-brown relieved with a few pale markings on the hindwings; the other has an all over black-and-white striped tiger-pattern. Both forms can occur among the offspring of a single female and both can occur in either sex. A single gene that switches development from one form to the other controls the difference.

A more common variation on this theme is the coexistence of two or more distinct forms of the female, while the males remain constant. Females of several of the familiar clouded yellow butterflies, genus *Colias*, such as the common sulphur, *C. philodice*, of North America (p.24), and the clouded yellow, *C. croceus*, of Europe, occur in two principal forms: bright yellow, like the males, or creamy white. The remarkable African mocker swallowtail, which can have several different forms of the female, all radically different from the male and distinct from each other, is discussed in the next chapter (pp.93–95).

RIGHT From beneath there is not much difference between male-like (top) and pale (bottom) females of *Colias croceus*, but from above the difference is obvious (compare *Colias* on p.24).

VARYING WITH THE SEASONS

In many parts of the world temperatures are sufficiently high to support successive generations of insects throughout the year. However, there is often a seasonal change in temperature and humidity, from hot and dry conditions to relatively cool and wet conditions. During wet periods the vegetation becomes green and lush, but at other times it is brown and parched. For any butterfly that relies on a degree of camouflage to survive, these changes are a challenge. To cope, some species can alter their development so that they produce adult butterflies of different colour, size or even shape, in anticipation of the different seasonal conditions into which they will emerge.

One of the most remarkable of these seasonally variable species is the African gaudy commodore, *Precis octavia*. In the cool, wet season in South Africa this butterfly is relatively small, mainly red above, has an even outline to its forewing and likes to hang around on hilltops. In the hotter dry season the butterflies are larger, mainly violet-blue above, more intricately patterned, have a wavy margin to the forewing, and seek out cool ravines, caves and even mine shafts.

Take 100 eggs from a female of either form of the gaudy commodore, divide them into two batches of 50, and rear one batch at 30 °C (86 °F) and the other at 16 °C (60.8 °F). The high temperature batch will develop into red, wet-season forms, while the low temperature batch will turn out as blue, dry-season morphs. They are using the temperature experienced during development as a cue to anticipate the likely coming season. They do not always get it right, however. In particular, if the caterpillars experience a temperature of about 24–28 °C (around 75–82 °F) during a critical stage in their growth, they emerge with an intermediate pattern. Between the seasons, butterflies of both forms can be seen roosting together and even mating.

In this case, although the ability to make such an amazing switch is genetically based, all individuals can produce the variations observed, dependent only on the environment in which they develop. They are not the result of individual genetic variation. Essentially similar, but less dramatic, seasonal changes affect many northern temperate species that have spring and summer generations. For example, spring individuals of the green-veined white, *Pieris napi*, have the veins on the undersides of the hindwings strongly marked with dark scales, but this is much fainter in the summer brood.

VARYING FROM PLACE TO PLACE

Conditions do not only vary with seasons, of course, but also from place to place – from lowlands to uplands, from savannah to forest, and from north to south. Butterflies with wide ranges, either ecologically and/or geographically, often show local variations. These variations can simply be environmental in origin, but most geographical variation probably has at least some genetic component. Sometimes these variations suggest the possibility of new species in the making.

RIGHT AND BELOW *Precis octavia*, the gaudy commodore, is a common African butterfly that regularly undergoes a remarkable change of appearance. In the dry season most individuals are bright blue above, with a row of red spots and a few dark marks.

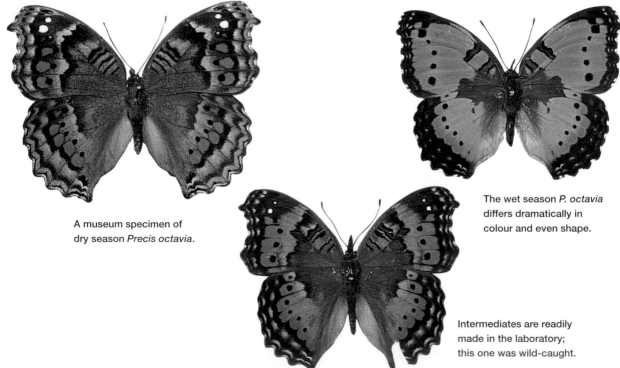

A museum specimen of dry season *Precis octavia*.

The wet season *P. octavia* differs dramatically in colour and even shape.

Intermediates are readily made in the laboratory; this one was wild-caught.

Parthenos sylvia virens from southern India.

P. sylvia lilacinus from the Malay Peninsula.

The more olive race from New Guinea, *P. sylvia guineensis*.

On Guadalcanal, *P. sylvia thesaurus* has a relatively dull coloration

The yellowish Sulawsesi race, *P. sylvia salentia*.

The clipper butterfly, *Parthenos sylvia*, ranges from Sri Lanka, India and Thailand through Indonesia to New Guinea and the Solomon Islands. The clipper is found in many countries and thousands of islands, and more than 30 local geographical races (subspecies) are currently recognized. Several of these are only separated on subtle, even uncertain or potentially unreliable grounds, but if we focus on the major changes that occur over the full extent of the clipper's range, the type of geographical variation seen in many widespread species can readily be appreciated.

In southern India the butterfly has a greenish-brown cast (race *virens*). Moving some 2,000 km (1,200 miles) to the east, in the Malay Peninsula, the local race *lilacinus* has a beautiful blue flush to the base of the wings. Another 2,000 km (1,200 miles) further east, on the island of Sulawesi, the race *salentia* is yellowish. In New Guinea (race *guineensis*) the wings are olive-brown. In the Pacific, on Gaudalcanal in the Solomon Islands (race *thesaurus*), the colours of the butterfly are duller still. These differences appear clear-cut, and indeed some of them are. But when we consider the butterflies from intervening areas, the separations may be much less obvious.

Across New Guinea, for example, the clipper varies only subtly from place to place. The differences are so small that previous attempts to divide the butterfly locally into a series of four or more discrete island races have now been abandoned. Disputes over the reality or otherwise of such divisions often exercise the minds (and even tempers) of butterfly taxonomists. One of the most remarkable cases of geographical variation in butterflies is the parallel variation seen in two South American passion vine butterflies, *Heliconius erato* and *H. melpomene*. This is discussed in the next chapter (pp.102–104).

ABOVE **Geographical variation between *Parthenos sylvia* races.**

CONTINUOUS VARIATION

The variation discussed so far involves discrete differences between the sexes, between seasons, from island to island and so on. So-called continuous variation also occurs. Size, such as stature in humans, is a good example, with classes such as 'tall', 'average' and 'short' not sharply separable, other than by making arbitrary divisions based on exact measurements. Butterflies vary in size in a similar way and this usually involves the interaction of 'nature' (genetic factors affecting size) and 'nurture' (notably the quality and quantity of food available to the caterpillars).

Characteristics such as colour can also vary continuously. In the European common blue, *Polyommatus icarus*, the males are always a lovely clear blue. The females are separable by a row of orange spots along the margins of the hindwings, but in addition they vary from dull brown right through almost every conceivable intermediate colour up to being almost as blue as the males. Here we have continuous variation in female colour, coupled with sexual dimorphism, and low variation in male colour.

BELOW In Ireland the female of the common blue, *Polyommatus icarus*, is large and often almost as blue as the male. In many areas of the UK such blue females are much less frequent, many being almost entirely brown.

ABOVE Six common blue specimens from England. Males (top left) are always bright blue. Females vary from almost as blue as the males (middle left) through intermediates to individuals that are almost completely brown (bottom right).

VARIATIONS IN CATERPILLARS AND CHRYSALIDS

It would be a mistake to imagine that only the winged, adult stage can vary. The larvae of many butterflies occur in multiple forms, such as those of the African large striped swordtail, *Graphium antheus*, which can be bright green, olive or purplish-brown. Chrysalids often vary in a similar way, the pupae of many swallowtail butterflies being either green or brown.

Rather like seasonal variation, in which the temperature or daylength cues switch on particular developmental pathways, the colour of the surface in which the caterpillars decide to pupate seems to act as a prompt. If the caterpillar pupates among green leaves, then the pupa it forms will usually be green, but if it chooses bark or dead foliage, then its pupa will normally be brown. However, among closely related species, some produce pupae in different colours while others only make pupae of one colour, regardless of where they pupate. This suggests that it is the *capacity* to produce chrysalids of different colours that is genetically controlled. However, it could also be that alternative genes affect both caterpillar behaviour and pupal colour: caterpillars destined to produce green pupae might instinctively seek out green leaves, and those fated to make brown ones might seek brown surfaces.

In many species there appears to be a combination of genetic predisposition and environmental flexibility. Experiments indicate that this system is only sensitive at a key stage. If a caterpillar sets out to pupate on a green leaf, and after the key stage is passed the leaf suddenly dies and turns brown, it will be too late: the pupa will be green, and unfortunately more conspicuous than would otherwise have been the case. This may be a powerful source of stabilising selection for 'getting it right'.

BELOW Chrysalids of the African orange dog or Christmas butterfly, *Papilio demodocus*, naturally occur in two different colour forms, green or brown. They usually match their background well.

VARIATIONS – THE MATERIAL BASIS OF EVOLUTION

The next chapter will look at the evolution of butterflies, and in particular the sorts of evolutionary change that can be brought about by 'natural selection'. In formulating their theory of evolution, Charles Darwin and Alfred Russel Wallace acknowledged that it was necessary not only to demonstrate that animals and plants varied in nature (not just in the laboratory or in the hands of livestock breeders), but also that such variations were at least partly heritable – as with many of the examples already discussed.

Individuals within a species can differ for all sorts of reasons – sometimes so strikingly that, at first sight, totally different species would seem to be involved (for example, female *Papilio dardanus* – p.94). In other cases, as with human blood groups, heritable differences, although hugely important, are totally invisible to the naked eye. Although small but visible differences are often just individual variation, seemingly trivial discrepancies can sometimes give reliable information for separating closely related species. One of the great challenges for biologists is to know which variations are individual, which are specific.

BELOW For nearly 250 years all zebra mosaic butterflies were thought to belong to a single species. But in 2001 systematists Keith Willmott, Luis Constantino and Jason Hall realised, based on larval ecology and behaviour, that two perfectly distinct South American species were involved – and demonstrated how their adults could be separated by particular small details in wing pattern formerly thought of as individual variation. *Colobura dirce* is on the left, *Colobura annulata* on the right.

CHAPTER 8
Evolving

ALL OF THE 20,000 OR MORE BUTTERFLY species alive today are presumed to have evolved from a single common ancestor. The age of this common ancestor is very uncertain. Perhaps it evolved 50 million years ago, the approximate age of the oldest fossilized butterfly that has so far been found (see p.9); almost certainly it was older, maybe much older than that. Yet we can draw some conclusions about evolutionary change by looking at the species around us today.

THE ORANGE OAKLEAF BUTTERFLY: NATURAL SELECTION AT WORK?

Sitting with wings closed, the orange oakleaf, *Kallima inachus*, is remarkable for its great resemblance to a dead leaf. By contrast, the uppersides of its wings are brightly coloured and conspicuous. Compare a number of oakleafs, all from the same place, and you will find that their upperside patterns are almost identical. Turn them over, however, and every one looks different. Some are like pale brown leaves, others darker or redder, some have transparent patches resembling rot-holes, or black marks that look like spots of mould. Why so constant on one side, but so variable on the other?

The upperside pattern probably plays a key role in communication among the butterflies, when a standard '*Kallima inachus* signal' will be advantageous, and is apparently under the influence of stabilizing selection. But to avoid detection when feeding on the forest floor, it is good to blend into the background of dead leaves. Dead leaves vary enormously in appearance, and it seems that every orange oakleaf can produce its own unique expression of the dead-leaf effect. If patterns were always the same, predators might soon learn that a particular sort of 'leaf' was always worth investigating. Selection that promotes variability is called 'disruptive' selection.

Regrettably, this species has never been investigated to find out exactly how the upper and underside patterns really do function. The ideas presented above are little more than educated guesses. Moreover, we have no information about how the variability of the underside is produced. It could be an interaction of different

OPPOSITE Two closely related tropical mapwing butterflies, *Cyrestis nivea* (two paler individuals) and *Cyrestis maenalis*, 'mud-puddling' on damp sand at Gunung Leuser National Park, Sumatra, Indonesia. Over the millions of years since the first butterfly appeared on Earth, what sorts of processes have shaped the differences between the species that now exist?

genes, or the result of an instability in the pattern-forming process. Despite 200 years of speculation, our knowledge of this wonderful insect remains rudimentary. Indeed, relatively few butterflies have been investigated scientifically to demonstrate if plausible evolutionary explanations are convincing or not. This chapter will look at some examples in which a degree of scientific rigour has been applied.

RIGHT The Indian race of the orange oakleaf butterfly, *Kallima inachus inachus,* shows off its brightly coloured upperside as it rests on a leaf.

RIGHT With its wings closed, the orange oakleaf looks just like a (dead) leaf. This camouflages the butterfly when it settles on the forest floor. When disturbed it may dive into a bush and fold its wings, and it will then be even more difficult to detect.

MIMICRY: SHEEP IN WOLVES' CLOTHING

On 26 April 1848, the naturalist and explorer Henry Walter Bates set out from Liverpool on an expedition to the Amazon. Bates had an excellent knowledge of butterfly anatomy and classification. For example, he knew that cabbage butterflies, Pieridae, walk on all six legs, whereas the nymphalids are 'tetrapods' and walk on only four legs, with their minute forelegs folded-up and held against the body. As he worked his way up the Great River, Bates discovered relatives of the cabbage butterflies which, instead of being the usual plain yellow or white, were brilliantly streaked in orange, yellow and black. When they were flying, these *Dismorphia* species were almost inseparable from certain, very common, nymphalids of the genus *Heliconius*, which had an almost identical design. As he moved up the river, Bates found that the colour patterns changed. But both sorts of butterfly changed in the same way, so that they were still confusingly similar to each other at any particular place. He had to check the anatomy of individual butterflies to tell them apart from each other.

Puzzled at first, the germs of the idea that would become his theory of protective mimicry began to form in his mind. The *Heliconius* were numerous, flew slowly, and behaved as if immune from attack. Bates reasoned that they must be unpalatable to potential enemies such as birds. Confident that they would be left alone, they could afford to be conspicuous. He next argued that if poorly defended species could somehow alter their appearance to become similar to the brightly coloured, protected species, then they could benefit. Predators might just pass over them, as more of the unpalatable sort. But how could a butterfly change its colour? Certainly not by an act of 'will'. And butterflies normally have offspring like themselves, so not just from one generation to the next.

On his return to England in 1859, Bates arrived to a storm of controversy surrounding publication of Darwin's *On the Origin of Species by Means of Natural Selection*. The Darwin/Wallace theory of evolution by natural selection gave Bates the clue that he needed – a mechanism by which changes in pattern could come about. Butterflies quite often produce individual variants. If an accidental change gave a slight resemblance to an unpalatable species, perhaps this might have a survival advantage. If the ability to produce the colour difference was passed on to its offspring, then they would inherit the same potential advantage. With variation in expression of the new colour pattern, some of the offspring would be more like the protected species than others, and have a better chance of avoiding attack. Here was something for natural selection to work on. At each successive generation, once the process had started, the appearance of the mimic would steadily be improved, eventually to reach the almost astonishing level of similarity between unrelated species that Bates had observed.

THE MONARCH AND THE VICEROY

ABOVE **The viceroy is closely related to various white admirals, black-and-white butterflies rather than orange. In a series of species hybridization experiments, biologist Austin Platt created a remarkable series of intermediates. Could mimicry have driven such a change in nature?**

RIGHT **Monarch larvae sometimes obtain enough heart poisons to kill a bird – and can pass these chemicals on to the adult. If a bird eats a poisonous monarch it will be sick, and remember the bad experience for months to come.**

Perhaps the two most studied butterflies in North America are the viceroy, *Limenitis archippus*, and the monarch, *Danaus plexippus*. As adults, the two species look remarkably similar, but they have very different eggs, larvae and pupae, and the caterpillars feed on unrelated plants. No other species of the genus *Limenitis* looks like the monarch, whereas the monarch is similar to other members of its own genus, *Danaus*. Perhaps we can infer that the viceroy changed from the normal appearance of the 'white admiral' genus to which it belongs to its current orange *Danaus*-like pattern. Could mimicry have brought about this transformation?

To demonstrate the idea that the viceroy is a mimic of the monarch, it would first be necessary to show that the monarch has immunity from predators. In the 1950s, entomologists Jane and Lincoln Brower started to investigate the chemistry of the monarch, by feeding adult butterflies to caged scrub jays. Shortly after eating just one butterfly, the birds were violently sick. Thereafter they avoided the brightly coloured monarchs on sight, and clearly remembered their traumatic experience even months

later. Further experiments suggested that a monarch could contain enough poison to kill a jay, so the vomiting was a protective reflex. The chemical concerned turned out to be a cardiac glycoside, a heart poison derived directly from the butterflies' milkweed host plants. Some milkweed plants that the butterflies feed on do not contain the glycosides, and monarchs reared on these plants will not make the birds vomit. But once a bird has experienced a poisonous monarch it will not risk sampling another one for many months to come.

The crucial test would be to see whether predators that had learnt to avoid the monarch would also avoid viceroys on sight. In most trials, the Browers' experiments delivered a clear 'yes'. But even though these laboratory experiments showed how, once established, mimicry might persist, it was also necessary to demonstrate that mimicry was powerful enough to bring about the initial evolutionary convergence. The theoretical work of statistician and evolutionary biologist Ronald Fisher during the 1920s suggested that a selective advantage of 1% for a beneficial change would have an appreciable effect on a population within a few hundred generations. By 'advantage' it is meant that those individuals carrying the altered gene responsible will have, on average, 1% more offspring than any that do not. If a gene changed the appearance of a non-mimic and gave such an advantage, it should spread fairly rapidly through the population. So the question then shifted to become one of degree – was it possible to measure the selective advantage of mimicry? (See 'Could mimicry change how species look?', on the next page.)

Bates published his theory in 1862, whereupon a huge debate erupted. Supporters of the new Darwinism hailed it as a wonderful example of evolution by natural selection, whereas antagonists sought every means possible to discredit the idea. For example, it was widely claimed that the birds did not eat butterflies. Observations could have been made and experiments devised to test these assertions, but the Amazon was too far away, and experimental biology was in its infancy. Even more critical, the development of statistics, now seen as essential for making sense of any relevant observations, was yet to come. The arguments about Bates' mimicry theory became largely an armchair exercise, without controlled investigations being made either in the laboratory or in nature. Critical experiments were not undertaken for nearly 100 years.

COULD MIMICRY CHANGE HOW SPECIES LOOK?

Callosamia promethea is a day-flying moth found widely in North America. The females are camouflaged and mostly inactive, resting in vegetation and attracting males by means of a powerful pheromone. The very active males are black, and considered to be mimics of a chemically protected swallowtail butterfly, *Battus philenor*. In the 1970s, entomologist Gilbert Waldbauer and his collaborators ran a series of remarkable field experiments in Illinois. They raised promethea moths in large numbers and divided the living males into three equal batches. All the individuals of one group were painted black and yellow, to resemble a palatable local swallowtail, *Papilio glaucus*. The second lot were painted orange and black to look like the ubiquitous monarch (or the viceroy). The third group they over-painted plain black, the natural colour of the moth. All the moths were then released simultaneously, a few kilometres from where virgin females had been placed in cages.

If selection by local birds were sufficient to maintain or promote mimicry, Waldbauer's team expected that fewer black-and-yellow painted moths would survive than those painted orange or black. Measured as the number of painted moths that found the caged females, the results were dramatic. Survival rates for the two sorts of 'mimics' were far better than that for the yellow *P. glaucus*-like form. On this evidence, selection in favour of a mimic could be massive – far greater than 1%.

THE AFRICAN MOCKER SWALLOWTAIL, *PAPILIO DARDANUS*

There are limitations to the effectiveness of Bates' idea of protective mimicry. For example, if a mimic becomes more numerous than the species it looks like, predators will come to associate the warning pattern with good food rather than disgust, to the potential disadvantage of both species. But as the promethea moth

RIGHT Three African milkweed species (left) and a male *Papilio dardanus* (bottom left). Three mimetic female forms of *P. dardanus* from mainland Africa (right) and a tailed, male-like female from Madagascar (bottom right).

experiments suggest, if a species could mimic more than one unpalatable species at a time (in this case a black mimic and an orange-and-black mimic), it might be able to sustain a larger population than if it produced only one warning pattern.

In eastern and central Africa, the common African mocker swallowtail, *Papilio dardanus* – once dubbed by evolutionary biologist Sir Edward Poulton as 'the most interesting butterfly in the world' – can have as many as five or even more different sorts of female, all flying in the same place. Each appears to be a good mimic of a different, chemically protected species. The yellow, tailed males are not mimetic. In some parts of the species' wide range, in Madagascar for example, and Somalia, the females have hindwing tails too, and all look very similar to the males. In parts of Ethiopia such male-like females occur together with mimetic forms.

Starting in the 1950s, zoologist Philip Sheppard and geneticist and physician Cyril Clarke investigated the genetics of this system. They discovered that inheritance of the female patterns seems to be under the control of a single gene, *H*, which occurs in multiple sorts, each one specifying one of the alternative patterns.

Both sexes carry *H*, but it never affects the male pattern. When an egg that will give rise to a female is fertilised by a sperm carrying a different version of *H* from the one it got from its mother, the developing female will only respond to one of the two versions. In this way the production of intermediate-looking patterns is prevented. Another gene controls the presence or absence of tails.

But how could a single gene account for all the coloration differences exhibited by the various female forms? *H* is almost certainly the site of a control or switch mechanism that can initiate different pathways during development, involving a whole series of other genes that only come into play once the process has started. Until recently this was little more than speculation, but an international team of scientists based in Europe, Africa and Australia has now located *H* and studied its DNA, confirming that it acts like a single 'regulatory' factor. Moreover, *H* is located close to several other genes that have been implicated in the control of wing coloration. This is a big step towards unravelling the mystery of how this remarkable butterfly can make so many different patterns.

Although our knowledge of the genetics and development of this fascinating butterfly is now rapidly improving, we still know relatively little in detail about its ecology or reproductive biology. However, there is absolutely no suggestion that the different female forms are new species in the making. Male mocker swallowtails can fertilize to find all of the different mimetic females equally well.

FALSE HEADS

Butterflies can use their patterns in other ways to avoid or trick would-be predators. False-headed butterflies are so-named because when they are sitting, wings folded, they appear to have a second head, with large eyes and long antennae, at the posterior end. This false head comprises a coloured lobe at the rear angle of the hindwing, with imitation eyes and antennae simulated by long, ribbon-like wing tails. The effect is usually enhanced by one or more radial lines on the underside of the wings that draw attention to the false head as a focal point. Although the real head is fully visible, the fake one at the other end is far more conspicuous.

Standing stock still on a narrow path in an Angolan forest I suddenly became aware of a false-headed butterfly, genus *Hypolycaena*, flying past at shoulder height. It settled on a leaf almost immediately in front of me. As it did so, it revealed the sheer cunning of the false head strategy. On landing, it flipped through 180°, to face in the opposite direction to the way in which

BELOW **Which way is it pointing? A false-headed butterfly,** *Hypolycaena lebona*, **from Sierra Leone. Note how the reddish stripe seems to draw attention to the dummy head.**

it had just been flying. Facing forward was the false head, with artificial antennae tantalisingly aflutter. Knowing how sensitive to movement any butterfly is likely to be, any effective strike should anticipate the expected direction of escape: onwards from where it has just come, and in the direction of where the head is pointing. So a strike is likely to be aimed just in front of the false head. When this happens, the butterfly shoots away in the opposite and totally unexpected direction. At best an attacker may grasp a few bits of tail, which the butterfly can quite well manage without – at least, compared with losing its real head.

FALSE EYES

If you are small and likely to be hunted, then suddenly being confronted by a huge pair of eyes is scary. If you are a small hunter, then you too may want to avoid large eyes, but small ones can be attractive as they may represent potential prey. Butterflies make use of both of these responses.

Disturb a peacock butterfly, *Aglais* (*Inachis*) *io*, at rest and it will suddenly open its wings to reveal its brilliant chestnut-red upperside and four great eyespots. Biologist David Blest investigated this butterfly experimentally in the 1950s and found that such a 'startle display' really did make young or inexperienced birds back off. Blest tried carefully removing the scales responsible for the eyespot pattern. The butterflies still behaved in the same way, but the startle response they produced was then much less marked. Mimicking the eyes of birds and mammals, with their pattern of concentric circles and highlights, evokes instinctive reactions.

BELOW At rest on a nettle, or overwintering with its wings tight shut (left), the underside of the peacock is almost black and rather leaf-like. If disturbed, the peacock can suddenly reveal its huge eyespots.

LEFT *Caligo* butterflies (owls) have one very large eye-spot on the underside of each hindwing. When spread out like a museum specimen, they do look a bit owl-like. But in life they always sit with their wings closed and upright above the body.

Perhaps the most remarkable, but probably least understood, eyespots belong to the South American 'owl' butterflies, in the genus *Caligo*. Two large and beautiful eye-like markings dominate the underside pattern of the hindwings, one on each wing. If a *Caligo* is 'set' as it would be for a collection, when the underside is looked at with the hindwings uppermost, the whole spread-eagled butterfly looks quite like an owl. Some have speculated that this owl-like aspect is an effective deterrent to predators. But it is unlikely that these insects, which mainly flap around in the dark forest understorey at dusk, would ever be seen in such a pose. By day the butterflies often rest on stems or tree trunks, or visit the forest floor to feed from fallen fruit. In both situations, with wings folded close together over their backs in normal butterfly fashion, only one eye can be seen at a time, and there is no owl-like appearance.

So if not 'owl mimicry', what are the huge eyespots for? According to Keith Brown, birds are very wary of them and will even stay away from a feeder if *Caligo* is settled beside it. The single visible eyespot may be large and realistic enough to suggest a lurking predator. Another possibility is that they help fend off lizards of the genus *Anolis*, which establish territories on tree trunks. Big *Anolis* are able to maintain larger territories, and relative eye-size may be used to settle disputes.

ABOVE American painted ladies, *Vanessa virginiensis*, on clover. Relatively small eyespots on the wing margins can deflect the attention of a predator from the body. Such areas can often be seen to have beak marks, where a butterfly has survived an attack.

So, if a *Caligo* resting on a tree trunk looks like a huge rival to an advancing lizard, then perhaps it will back away. Far fetched? Perhaps, but at least this is a testable idea, first proposed by the late David Stradling, English entomologist and conservationist.

Small concentric eyespots are a major feature of many nymphalid butterflies, perhaps most notably among the 'browns' (Satyrinae). These spots are always ranged along the outer margins of the wings, never close to the body. When a bird attacks a butterfly it tends to peck at the eyespots. By having them near the edges of the wings, the deadly beak is kept away from the butterfly's body. The cost of a few tears to the wing edges is a small price to pay if it saves your life.

FLASH AND DAZZLE

Part of the startle effect of the peacock does not depend on the big eyespots, but just on the great contrast between its appearance one moment and the next: dull and drab with its wings closed, then bright and dazzling with them open. Simply by opening and closing its wings sharply a few times, these rapid changes in appearance can be enough to startle a predator, or at least make it hesitate long enough for the butterfly to escape. Behavioural experiments have shown that young and relatively inexperienced birds suffer from what has been termed 'neophobia': the fear of something new or unexpected.

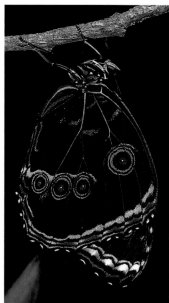

Possible confusion caused by a change in appearance may also help butterflies in flight. Brilliant blue *Morpho* butterflies of South America are mirror-like above, but dark beneath. As they fly along a trail or follow a river through the dark forest, each wing-flap can take the butterfly several metres forward. Viewed from the side, the alternation of dazzling blue to near invisibility can make the flight path of the butterfly very difficult to follow. A bird or other predator trying to strike at it in flight might have a very hard task to anticipate the trajectory.

CAMOUFLAGE

So far we have talked about defensive schemes that work when the butterfly has been detected. There are alternatives, one of which is to try to avoid being recognised at all, even if in full view. This is the strategy of camouflage or *crypsis*. Among European butterflies, the beautiful green underside of the green hairstreak, *Callophrys rubi*, makes it very difficult to spot when it is settled amongst leaves, and likewise the mottled underside pattern of the grayling, *Hipparchia semele*, renders it almost invisible when at rest on a bare patch of soil. However, the camouflage that excites our sense of wonder most of all is so-called 'leaf mimicry'. Some butterflies at rest appear remarkably like dead leaves, with simulated midribs, veins and even rot holes. The most famous of these is the orange oakleaf, *Kallima inachus*, already mentioned (p.90). The leaf shape acts as a disguise when the butterflies are feeding on the ground, surrounded by fallen leaves.

A less obvious but fascinating example that works amongst green leaves is found in the genus *Gonepteryx,* the brimstone butterflies of temperate Europe and Asia. Viewed in isolation from leaves, their undersides seem to have no more than

ABOVE *Morpho achilles* rests on a leaf, exposing its dazzling blue upperside (left). But seen with wings folded (above), it presents a totally different appearance. These butterflies often feed from fruit on the forest floor, when this pattern offers good camouflage, perhaps with intimidating eyespots for good measure, if a potential predator gets too close. When a morpho flaps along dark jungle paths and streams, the alternate on–off flash of brilliant blue and black can make the butterfly very difficult to follow.

a passing resemblance to leaves. Being pale lemon yellow rather than green, they seem to be the wrong colour. However, as discussed by Russian entomologist Boris Schwanwitsch, they exhibit a poorly understood phenomenon that he called the 'green reflex'. When sitting amongst foliage, the wing undersides reflect or take on the green colour of the leaves surrounding them and then the butterflies become very difficult to see – another wonderful example of camouflage.

RIGHT This grayling, *Hipparchia semele*, looks fairly obvious against grass, but these butterflies usually rest on patches of bare earth. With their frequent habit of tilting their wings towards the sun, perhaps to eliminate shadow, they literally seem to melt into the background.

RIGHT Close up, this brimstone, *Gonepteryx rhamni*, also looks fairly obvious. However, the 'green-reflex' ensures that the leaf-like shape takes on a greenish hue very similar to any leaves that the butterfly rests amongst – making it very difficult to see from a distance.

TRICKS THAT CATERPILLARS PLAY

In the course of evolution the highly vulnerable larvae have evolved a range of strategies to outwit potential enemies. Unlike chameleons, caterpillars are not able to change appearance at will, but some undergo two or even three radical changes in colour, pattern, ornamentation and even shape as they grow. Best known for this are the swallowtails, in which the middle larval instars are dark, with irregular white marks across the middle of their body. Rather than make the caterpillars conspicuous, this makes them look like bird-droppings, objects of no interest to predators as the larvae rest motionless on the surface of a leaf during the day. On reaching the final instar there is usually a dramatic alteration. Apparently too big now to make a convincing bird-dropping, or perhaps too busy feeding to stand still, these larvae usually change to a beautiful glossy green with a dark 'saddle' mark. The green is inconspicuous amongst leaves, and the saddle breaks up the characteristic caterpillar outline.

An awful lot of what caterpillars eat is relatively indigestible, and they make a lot of dung. Healthy caterpillar faeces, which go by the wonderful name of frass, are more like guinea-pig pellets than cow dung. Some of the caterpillars' deadly enemies, the parasitic wasps, home in by detecting the characteristic odour of frass pellets, which often collect on the leaves immediately below where a caterpillar is feeding. To overcome this, the larvae of some species have developed the extraordinary trick of explosive defecation. At just the right moment the caterpillar gives a smart twitch of the end of its body, flicking the frass pellet 20–30 cm (8–12 in) away so that it falls to the ground, too far away to give any wasp an accurate 'fix'. The selective advantage of any trick that can defeat parasitic wasps must be very great.

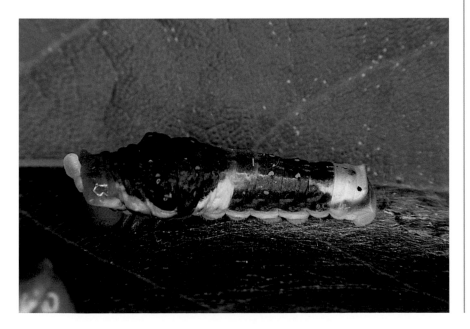

LEFT **This swallowtail caterpillar, *Papilio*, could easily be mistaken for a bird-dropping. The larvae of many tropical swallowtails share this appearance in their middle instars, but usually change to green and adopt another pattern after the final moult.**

Another way to avoid predators is by feeding inside a plant rather on the surface. Many blue butterflies feed inside seeds and flowers, although the primary need here appears to be high-quality nutrition rather than defence. The American giant skippers feed inside the leaves and stems of their host plants (p.46), but such internal feeding is unusual amongst the butterflies. Instead, many species use silk to sew leaves together to make tubes or shelters. They may emerge from these to feed at night, or eat the leaves of the shelter from inside and then move on. However, certain birds seem well aware of this trick and will systematically open sewn-up leaves, apparently confident they will find something good inside. No ruse is ever 100% effective for long, and so the evolutionary 'arms race' will always continue.

MIMICRY AND GEOGRAPHICAL VARIATION

We have seen that it is possible for natural selection, depending on circumstances, to keep features constant, bring about changes in appearance, or promote variability. But how does diversification of species come about? Can natural selection bring about the evolution of new species, as Darwin suggested?

Perhaps the most remarkable case of geographical variation in butterflies is the parallel variation seen in two South American passion vine butterflies, *Heliconius erato* and *H. melpomene*. Within the genus *Heliconius*, these two species are not closely related, but they occur together across most of tropical Latin America. Wherever they are found together they look remarkably alike, but from place to place they look amazingly different. The explanation for this local similarity is considered to be mimicry, but of a different sort to that proposed by Bates (see p.91). These butterflies are capable of producing cyanide. German biologist Fritz Müller, following Bates, suggested that convergence in pattern between protected species would also be advantageous, if each new generation of predators had to attack butterflies to learn the significance of particular bright patterns.

But why do the shared patterns change from one area to the next? This divergence across regions appears to be due to historical differences in the mix of *Heliconius* and other chemically defended butterflies and moths, leading to local differences in pattern. At first sight you might imagine that each region was inhabited by one very distinct species of *Heliconius*. Instead, we now understand that the two species shift their colour patterns in parallel, from one zone to the next. Again, we must ask, is there any experimental evidence that these colour patterns are effective? Biologist Woody Benson painted Costa Rican *H. erato* to look like *H. erato* from other areas. The result was again consistent with mimicry as a very powerful selective agent able to maintain, or even bring about, dramatic changes in colour pattern.

HELICONIUS MELPOMENE AND *H. CYDNO*

When Bates first encountered such regional shifts in mimetic patterns, he speculated that this might give a clue to how new species could be formed. However, since the discovery that the remarkably different-looking races of *H. melpomene* are all inter-fertile (able to breed with one another), and that the same applies to *H. erato*, mimicry amongst these butterflies is seen to be about the evolution of colour patterns *within* species, not the formation of new species. However, the introduction of molecular techniques recently identified the strong possibility that, in some cases at least, the evolution of mimicry (and thus natural selection) has, after all, been involved in species formation.

The closest relative of *H. melpomene* is *H. cydno*, a butterfly that looks like yet another *Heliconius*, *H. sapho*. These two share very similar black-and-white or pale yellow patterns, and never look like *H. melpomene* or *H. erato*. Molecular data indicate that *H. cydno* is very closely related to *H. melpomene*, and they occasionally form hybrids in nature. Although the hybrid females are sterile, the males are fertile and can cross with either parental species. However, such hybrids are poorly adapted, as they do not conform to the warning signal pattern of either group, and will thus be selected against. Mate choice experiments by zoologists Chris Jiggins, Jim Mallet and colleagues have now shown that the colour patterns not only function in mimicry, but are also used by the butterflies to tell each other apart, and generally avoid courting and mating with the wrong species. However, when *H. melpomene* and *H. cydno* from different areas are brought together, they are much more likely to make a mistake.

This example implies that the evolution of new mimetic forms can lead, through mating preferences, to the separation of new species. Indeed, some butterflies, such as those in the remarkable African genus *Pseudacraea*, appear to have what the late Miriam Rothschild would have called 'a gene for mimicry': each one of several different species mimics a different chemically protected butterfly or moth. So it appears that there can indeed be a link between natural selection and the origin of species, as Henry Walter Bates seemed to sense as he explored the wonders of the Amazon valley.

GETTING IT WRONG – A ROLE FOR CHANCE

One of the most peculiar host plant deviations in the butterflies occurs amongst the South American satyrines. Most species of *Euptychia* feed on grasses, but a few develop on *Selaginella*, a club moss that often grows amongst grasses. The clubmosses are an ancient group of non-flowering plants, and it seems certain that this host change must be recent, in an evolutionary timescale. In Southeast Asia another genus of browns, *Ragadia*, has made the same shift. The larvae of both groups look similar and differ markedly from normal satyrine caterpillars, being beautifully camouflaged against the branched clubmoss filaments. A comparison of *Euptychia* and *Ragadia* would make a fascinating study concerning the consequences of such a radical host change. But the *origin* of these shifts seems to be due to chance: once in an evolutionary while, larvae can survive on other plants, and may end up getting 'hooked'.

PUTTING IT RIGHT – A ROLE FOR INDIVIDUAL ADAPTABILITY

Much of this chapter has focused on natural selection acting on heritable variations to explain how butterflies look today, after millions of years of evolution. Most of the examples are about coloration, as it is relatively easy

to appreciate how, when faced with visually hunting predators like birds and lizards, natural selection could have wrought adaptations such as mimicry and crypsis.

However, it is all too easy to imagine that the evolutionary process is driven solely by accidental genetic changes selected as a result of environmental challenges – such as climate shifts, new predators, or diseases. Like all organisms, butterflies are also individual agents in their own right, in control, to some degree, of their own destiny. Adaptive behavioural changes can often help 'put things right' or overcome limitations that might otherwise place the organism at a disadvantage. Thus, for example, flexibility in development in response to varying conditions, coupled with behavioural adjustments, can offset survival problems caused by seasonal changes (p.81).

The ability of individual organisms to adapt to circumstances as part of the fundamental process of living is considered by some evolutionary biologists to have significant consequences for the processes of evolution. Even though natural selection acting on whole populations is still seen as very important, the actual course of evolution appears to be shaped by a number of interacting factors – including individual adaptability.

ABOVE In the South American genus *Euptychia*, the switch from 'normal' grass or sedge feeding (left) to very unusual club moss feeding (right) has been marked by a striking change in appearance. Camouflage appears to be the explanation. Remarkably, a similar, convergent evolutionary change has affected the caterpillars of the distantly related *Ragadia* butterflies of Asia, which also feed on clubmosses.

CONTINUING EVOLUTION

The study of evolution inevitably involves much speculation and inference, and it would be a mistake to imagine that we understand the course of evolution in butterflies. All we can say is that the various studies made so far are largely consistent with the idea that evolution progresses through a continual interaction of behaviour, accident, environmental change and selection. According to ecologist Mike Singer, like all organisms, butterflies are just good enough to get by. Given that nature is so dynamic, and that butterflies have so much potential for variation, we can be confident they have evolved in the past, and will go on evolving in the future. They will certainly need to, if they are to survive the challenge of a world dominated by humans.

CHAPTER 9
Butterfly futures

WHAT IS THE FUTURE FOR BUTTERFLIES? Faced with pollution, climate change and habitat destruction, will they survive to delight, intrigue and fascinate future generations, or will they just disappear? There is little doubt that at least *some* butterflies will survive all but the most extreme changes that we can anticipate. But if current trends continue, many butterfly species, and a far higher proportion of subspecies and populations, will face extinction over the next 100 years.

ADAPT AND SURVIVE

In 1997, near Lewes in Sussex, a butterfly never seen flying in England before was found breeding in a garden. This African species, *Cacyreus marshalli*, was seen again in Sussex the following year. Was this sudden arrival evidence of global warming? The butterfly's common name, the geranium bronze, gives a clue. A native of southern Africa, its larvae develop on wild and cultivated geraniums, feeding inside buds and seeds as well as on leaves. With the expansion of

OPPOSITE Fivebar swordtail, *Graphium antiphates* (foreground) and common jay, *Graphium doson*, at a river bank in Gunung Mulu National Park, Malaysia. Like so many butterflies, the wonderful swordtails are dependent on forests for their survival – but forests are being cut down for fuel, timber, agriculture and 'development' almost everywhere. Some butterflies have already been driven to extinction – what does the future hold for the rest?

LEFT An invader: *Cacyreus marshalli* was recently introduced to Europe from Africa as a result of the horticultural trade of its geranium host plants.

ABOVE **The holly blue, *Celastrina argiolus*, is masterly at finding scattered places where the larval foodplants, including holly and ivy, grow.**

horticultural trade between Africa and Europe, larvae of the geranium bronze have been accidentally transported on a number of occasions. It is now spreading across various Mediterranean countries, and has become a pest to geranium growers in Majorca and elsewhere. The geranium bronze is spreading because, once introduced, it happens to be well adapted to human environments, such as parks and gardens. But there are other characteristics that mark out butterflies that are destined to survive – and those that will fail.

CHARACTERISTICS OF GOOD SURVIVORS

Dutch ecologist Fritz Bink noted that although many butterflies are declining in north-western Europe, some are thriving and actually increasing in numbers and range. He endeavoured to show that particular biological characteristics could be linked with good survivorship. He used these characteristics to identify what he called 'the butterflies of the future'. According to Bink, such temperate region butterflies usually overwinter as adults or pupae, have well-developed dispersal ability and produce two or more generations per year. Nomads, they are able to find and breed in almost any patch that offers the necessary resources. In southern UK, the holly blue, *Celastrina argiolus*, overwinters as a chrysalis, the adults roam freely and it has two generations per year. Found throughout North Africa and Southeast Asia, as well as the whole of the north-temperate region from Ireland to Japan, the holly blue is surely a butterfly for the future. Even so, its numbers can fluctuate quite widely from year to year.

CHARACTERISTICS OF POOR SURVIVORS

Bink's work not only revealed what type of butterfly could survive in the fragmented and ever-changing European landscape, but also identified the characteristics of 'failures'. These are species that can only survive in particular habitats (Bink termed them 'fixtures'), usually overwinter as caterpillars and have slow growth rates or only produce one generation per year. These biological features are typical of the 17 species of butterfly that have become extinct in Holland over the last 300 years. Likewise, many tropical butterflies are threatened because they are 'fixtures' or have small ranges – and the tropics are where most species of butterflies occur.

Wholesale destruction of certain habitats, notably the clear-cutting of former continuous rainforests, is likely to have the greatest impact. For example, of the approximately 850 species of butterfly found in the Philippines, more than one-third are found nowhere else. Many of these are restricted to just a few of the islands, and perhaps 50% are specialists unable to survive without mature rainforest. As deforestation of the Philippines moves to near totality over the next 50 years, we must anticipate that more than 150 species of the world's butterflies are likely to disappear forever. Repeat this across the world, and we have to acknowledge that the future for many butterflies looks bleak.

IS CLIMATE CHANGE HAVING AN IMPACT?

Although butterflies are primarily tropical, specialized species occur at all latitudes. In North America there are several browns, including 'alpines' (*Erebia*) and 'arctics' (*Oenis*), restricted to Alaska or parts of northern Canada. On the other side of the globe, three distantly related browns (*Oreixenica* and *Nesoxenica* species) are found only in Tasmania. In the face of global warming, these butterflies would probably retreat higher into the mountains. But if they go higher, they will have to depend on less and less land and their chances of local extinction in unfavourable years will increase dramatically. Is there any evidence that this could really happen?

Even though the extent of global warming due to increasing carbon dioxide in the atmosphere remains debatable, records made over the past 100 years show that the climate of western Europe has warmed by at least 0.8 °C (1.4 °F). This has led to a northern shift in temperature zones of about 120 km (75 miles). American ecologist Camille Parmesan, in collaboration with 12 European experts, demonstrated that even this seemingly small change has had a measurable effect. Out of 35 non-migratory western European butterfly species, three showed some evidence of a southern shift, but 22 had undergone polar shifts of 35   240 km (22–149 miles) at their northern boundaries.

Some predictions have suggested a temperature rise in northern Europe of more than 4 °C (7.2 °F) over the next 100 years, so the impact of global warming on cool-temperate species could be dramatic – and in some cases probably very unwelcome. On the one hand, species restricted to isolated high mountains could find themselves literally driven upwards to extinction, as their habitats shrink and even vanish altogether. On the other hand, we can expect to see some local additions. Based on current trends, it was predicted some years ago that Scotland would gain species such as the small skipper, *Thymelicus sylvestris*, which was forecast to reach Edinburgh by 2012, and the large skipper, *Ochlodes venata*, which was predicted to arrive in Glasgow by 2020. By 2014 both had colonized Scotland, with the latter as far north as mid-Ayrshire, but the former was still 50 km (30 miles) short of Edinburgh.

BELOW *Erebia youngi* on Eagle Summit. This mountain ringlet butterfly is restricted to dry tundra and scree slopes in western Alaska and the Yukon. Global warming is a direct threat to the existence of this cold-adapted insect.

HUMAN EFFECTS ON HABITAT CHANGE

Temperature is not the only cause of habitat change. Another is the ever-changing pattern of human land-use. In the UK and many other parts of Europe, the traditional rotational woodcutting system of coppicing produced a fine mosaic of open and regenerating woodland patches that provided ideal conditions for a number of butterfly species, such as the heath fritillary, *Mellicta athalia*, and chequered skipper, *Carterocephalus palaemon*. Commercial coppicing has now been almost abandoned in the UK, and the chequered skipper became locally extinct in England in 1976, apparently as a result. Survival of the heath fritillary in some areas is only ensured by the success of conservationists, who have persuaded a few landowners to continue with the traditional form of woodland management.

More generally, habitat change has a huge impact, mainly through breaking up former continuous areas of suitable territory into patches or small fragments. Industrialised agriculture, with the aid of herbicides and pesticides, encourages monocultures unsuitable for all but a few species.

BELOW A chequered skipper, *Carterocephalus palaemon*, photographed in 1996 near Fort William. This butterfly became extinct in England in 1976, but since about 1939 some 40 colonies have been discovered in the western highlands of Scotland.

THE EFFECTS OF POLLUTION

Surprisingly little is known about the effects of pollution. Although the widespread decline of butterflies seen in many rural areas of North America, Europe and Japan is often attributed to the use of pesticides, there is little evidence that these chemicals have a direct effect. Indeed, one classic study on the use of chemicals against small white butterflies suggested that insecticides made the pest problem worse. This was because the pesticide killed off one of the butterflies' most potent natural enemies, a parasitic wasp, more effectively than the caterpillars, which were protected by the layers of cabbage leaves. Fortunately, most insecticides are used on crop monocultures or in horticulture and not on butterfly food plants. One exception may be field margins, where spray drift can affect those butterflies adapted to such open weedy patches.

The use of herbicides may have a more serious, but indirect, effect. The regular application of plant fertilisers used in intensive agriculture encourages a very vigorous growth of crops, completely excluding the wild flowers and other small herbs on which so many butterflies depend. Airborne pollution, such as particles that clog plant leaves, and sulphur dioxide, can also have a serious impact on wildlife, mainly through acidification and poisonous heavy metals. Again, however, there is little evidence of a direct effect on butterflies. The impact seems indirect, for example favouring coarse grasses at the expense of broad-leafed herbs and so encouraging common browns while discouraging rare fritillaries.

A new form of pollution, according to some opinion, is the use of genetically modified crops such as *Bt* corn. The letters *Bt* stand for *Bacillus thuringiensis*, a naturally occurring bacterium. Some strains are lethal if eaten by the larvae of almost any moth or butterfly, due to a specific protein. Preparations, available as a powder or spray, are used widely in horticulture and forestry. Genetic engineering has now been used to add the DNA responsible for making the lethal bacterial protein directly to the genome of a number of crops, including corn. Any caterpillars eating these crops, such as those of corn-borer moths, will die. At first sight this technology appears to be good as there is no need to spray with indiscriminate pesticides.

In the USA corn is grown on more than 70 million hectares (over 170 million acres) of land and is a crop that not only feeds American citizens, but also provides huge surpluses capable of sustaining famine relief programmes around the world. One-quarter of this output is now produced using *Bt* corn. Are there avoidable or unacceptable costs of this to wildlife? In 1999, John E. Losey and colleagues at Cornell University in New York State published an article in the international science journal *Nature* claiming that transgenic *Bt* corn pollen harms monarch larvae. As the genetically modified corn ripens, massive quantities of pollen, carrying the *Bt* protein, are carried on the wind over surrounding areas. Here the tiny grains fall on the leaves of wild plants and can be eaten by caterpillars. The Cornell researchers found that monarch larvae suffer from ingesting *Bt* corn pollen, with a reduced growth rate and increased mortality. Consequences for the ability of survivors to reproduce and migrate successfully are as yet unknown. Similar results have been claimed from laboratory trials with the European swallowtail, *Papilio machaon*. Not surprisingly, corn seed manufacturers have hotly contested such laboratory results and their possible significance. The debate continues.

REAL LOSSES

The Indian Ocean island of Mauritius is infamous for once being the home of the dodo. This remarkable flightless bird became extinct by the mid-seventeenth century due to over-hunting by humans and the introduction of rats. Less well known is that the Mauritian mother-of-pearl butterfly, *Salamis augustina vinsoni*, has also become extinct since the arrival of European settlers. However, what is considered to be a local variant of the same species still survives 250 km (155 miles) closer to Madagascar, on the less deforested island of La Réunion.

The most famous of all extinct butterflies is the xerces blue. This little butterfly was native to the coastal part of California that has become San Francisco. The last xerces blue ever seen was collected on 23 March 1943. Some specialists regarded the xerces as a separate species, but most now consider it to be a subspecies of the silvery blue, *Glaucopsyche lygdamus*, a locally variable butterfly still widespread throughout most of North America. But the relentless 'development' of the xerces blue's habitat struck a chord with naturalists, and the world's first organisation devoted to the conservation of invertebrate animals was born: the Xerces Society.

BELOW **The last specimen of the Mauritian mother-of-pearl butterfly,** *Salamis augustina vinsoni*, **ever seen, collected in 1957. Now considered locally extinct, one remaining population of this insect, a separate subspecies, survives on the neighbouring island of La Réunion.**

ABOVE **Is the xerces blue extinct?** Undoubtedly the San Francisco Bay population, with its characteristic underside (top right) has gone, overwhelmed by developers. But *Glaucopsyche lygdamus* (bottom) survives elsewhere in a series of subspecies, among which the xerces was probably just one.

The mother-of-pearl and the xerces blue tell us something about the extinction of butterflies. So far there have been very few losses of whole species. Instead, we see 'death by a thousand cuts', as one local population after another disappears. Some butterflies, like the holly blue, recolonise suitable areas easily, but far more are like Bink's fixtures, or have low colonization rates. For most of these species all is not yet lost, but action, as with the heath fritillary, will be needed for more and more of them as the areas they naturally occupy get smaller and smaller. Relatively small regions with many unique species, such as Cuba, Madagascar, peninsular India, the Philippines, Sulawesi and New Guinea, are also a particular cause for concern.

WHAT CAN BE DONE TO HELP?

In Europe as a whole, more than 10% of all resident butterfly species are now threatened with extinction. In such developed, temperate countries with relatively small numbers of species, recovery plans for individual butterflies can be made.

Even so, because resources are always limited, such plans tend to be put in place only when the decline has reached a near-critical stage. Moreover, to be successful such plans usually depend on a very deep understanding of the species' precise ecological needs. Research takes money and, crucially, time. By the time it is recognised that a particular species is locally or regionally endangered, essential research is done, and an effective plan is worked out, it may be too late.

Four out of the 59 resident butterfly species found in the UK have become extinct over the past 150 years. In recent decades several more have undergone marked or even dramatic decline and are likely to go extinct in the UK unless something is done to help. One of these is the heath fritillary, already mentioned above. In 1980 it became apparent that this butterfly was headed for extinction in England unless an effective recovery plan was implemented. In the East Kent part of its UK range, once its dependence on coppiced woodland was realised, the necessary action became clear, and this has been carried out very successfully. Between 1984 and 1998 the butterfly was reintroduced to four woods in Essex. However, as a result of intensive research post-1980, it became apparent that it was also still present but declining in a few localities in south-west England, notably on Exmoor, and in parts of Devon and Cornwall. Despite intervention, decline continued in the south-west until very recently. It proved necessary to get an intimate understanding of the local conditions – what had been learnt and tried successfully in Kent, only 300 km (190 miles) away, did not give all the answers.

HABITAT PROTECTION

In most regions of the Earth, many species are dependent on one or more particular types of habitat. Certain habitats coincide with areas attractive for development, such as housing or agriculture. In the USA the loss of prairie grasslands for farming is almost complete. In Brazil most of the Atlantic Forest has been cut down. In Asia huge areas of mangrove have been destroyed. In Australia much of the native mallee scrub has been transformed for cereal crops. Each original habitat had its own suite of species. These assemblages are now mostly lost, with the individual species surviving here and there in habitat fragments, at a high risk of local or even global extinction. It is therefore necessary to protect representative samples of such threatened habitats for many species to have any realistic chance of long-term survival. In a few cases it may be possible to restore entire habitats by careful management, building out from pockets or remnant patches of the original ecosystem, as biologist and conservationist Dan Janzen has done so successfully in the Guanacaste dry forests of Costa Rica. A further refinement of the habitat approach is to work at the bioregional or landscape level, where a variety of different habitats, each supporting a different mix of butterflies and other species, are managed for conservation within an integrated, whole system. This has many advantages, but is likely to be expensive and complex, as so many

ABOVE Swedish, but alive and well in western England. Photographed in July 2002, this *Maculinea arion* represents a successful reintroduction of a locally extinct butterfly. It took a huge effort that could only be matched for a handful of species.

different stakeholders and agencies have to agree if effective action is to be taken.

At an even more general level, it may be necessary to set up conservation area networks, identifying a range of different places (estates, habitats, landscapes) where one or other form of conservation action is needed. Ideally such networks address priorities and have stated goals for conservation that can be measured and monitored over time to ensure that they are really working as intended.

If all else fails, so long as there are suitable populations to be found somewhere, a species can be reintroduced to an area where it lived formerly (as in the case of the heath fritillary in Essex). This can be seen as an individual species recovery or action plan, or it can be done as part of a habitat restoration plan, or even a landscape project. Perhaps the most celebrated case of successful reintroduction has been the re-establishment of the large blue, *Maculinea arion*, in England, using butterflies from southern Sweden.

SHOULD YOU CARE?

Is there any reason to care about butterflies? Does it matter if some, which most of us never see in a lifetime, go extinct? Does it matter if they all go extinct? Who cares if it means more food, better housing, less poverty? Some butterflies, like the lime butterfly, *Papilio demoleus*, and the cabbage white, *Pieris rapae*, are pests. A few are important in the pollination of wild flowers, but why care about them either? Some butterflies are directly useful, like the larvae of the semi-domesticated pierid *Eucheira socialis*, which are eaten in Mexico. In Africa, as noted by Tanzanian entomologist Steven Liseki, the appearance of certain species forewarn farmers of rain, and tell them that it is time to go to the fields.

These utilitarian values are but pinpricks. The real value of butterflies, if there is any, lies in their cultural significance, in what they can tell us about our own environment, and for some people at least, in their very existence. Butterflies appear on Egyptian tombs, symbols of metamorphosis and the after-life. Morpho butterflies are seen by certain South American Indians as evil spirits. Migrant monarchs are seen by Mexican poet Homero Aridjis as returning souls of the dead. In Japan the magnificent emperor butterfly, *Sasakia charonda*, is celebrated as a national symbol. More prosaically, butterflies can be used to monitor the 'health' of an ecosystem. As leading environmentalist and butterfly specialist Paul Ehrlich has long pointed out, a world that cannot sustain butterflies is unlikely to sustain us either.

Personally, I make no pretence of the fact that I just love butterflies and that my whole life has been enriched by trying to get to know them. The thought of their extinction, even those that I will never see or know, is painful – and I am confident that I am not alone in such a view, that nature is intrinsically valuable. Biologist Ed Wilson has called this human love of nature 'biophilia'.

TAKE ACTION

As an individual you can do a lot to help conserve butterflies, either directly or indirectly. Those with the time and energy can join local conservation and entomological societies. These organisations will often know about or be involved in local activity, where you can help through practical field work, fundraising or offering administrative support. Funds and conservation action may be targeted at particular species, habitats or landscapes.

Less directly, you can set a good example by adopting green policies in your own home, by reducing your use of fossil fuel, for example walking instead of taking the car, and voting for local and national parties or individual politicians known to be active on environmental issues. Get involved with like-minded people, adopt a species, adopt a habitat, try not to pollute, encourage green politics. As a citizen, as a volunteer, as a scientist, as a lover of nature, you can have a real impact on the future of butterflies.

Appendix

Family/subfamily	Common name	Approx. number of species	Natural geographical spread
Family Papilionidae	apollos, swallowtails, birdwings	550	worldwide except New Zealand
Subfamily Baroniinae	baronia, or archaic swallowtail	1	Mexico
Parnassiinae	apollos	70	north temperate and Himalayas
Papilioninae	swallowtails and birdwings	480	worldwide except New Zealand
Family Hedylidae	American moth butterflies	40	Caribbean, Central and South America
Family Hesperiidae	skippers	4,000	worldwide
Subfamily Coeliadinae	awls, policemen	75	Africa and Indo-Australia
Euschemoninae	regent skipper	1	eastern Australia
Eudaminae	longtails	400	mainly Americas
Pyrginae	spread-wings, flats, firetips	800	worldwide but majority South America
Trapezitinae	Australian skippers	60	Australia and New Guinea region
Heteropterinae	chequered skippers, skipperlings	150	Americas and north temperate
Hesperiinae	swifts, grass skippers	2,500	worldwide
Family Pieridae	whites and sulphurs	1,000	worldwide
Subfamily Dismorphiinae	wood whites, mimic sulphurs	100	South America and north temperate
Pseudopontiinae	frail whites	5	African rainforests
Coliadinae	sulphurs and yellows	250	worldwide except New Zealand
Pierinae	whites and orange tips	700	worldwide

Family/subfamily	Common name	Approx. number of species	Natural geographical spread
Family Lycaenidae	blues, coppers and hairstreaks	6,200	worldwide
Subfamily Curetinae	sunbeams	20	tropical Asia
Aphnaeinae	silver-lines, African coppers	250	mainly Africa, rest tropical Asia and Middle East
Poritiinae	rocksitters, buffs and gems	700	Africa and tropical Asia
Liphyrinae	moth butterflies	35	Africa and Indo-Australia
Lachnocneminae	woolly-legs	40	Africa
Miletinae	harvesters and brownies	100	mainly Africa and Indo-Australia
Lycaeninae	coppers	100	N. temperate, Africa, New Guinea, New Zealand
Theclinae	hairstreaks	3,000	worldwide
Polyommatinae	blues	2,000	worldwide
Family Riodinidae	metal-marks	1,400	mainly New World
Subfamily Euselasiinae	metal-marks	200	South and Central America
Nemeobiinae	punches and judies	100	Old World
Riodininae	metal-marks	1,100	mainly South America
Family Nymphalidae	brush-footed butterflies	7,000	worldwide
Subfamily Libytheinae	snouts (or beaks)	12	almost worldwide
Danainae	milkweeds, glasswings	500	worldwide, mainly tropics
Calinaginae	freaks	10	Himalayas to China, Taiwan, Thailand, Vietnam
Charaxinae	rajahs, pashas	500	tropics extending into temperate
Satyrinae	browns, morphos, owls	3,500	worldwide
Limenitidinae	admirals, viceroy, sisters, sailers	600	almost worldwide
Heliconiinae	fritillaries, yeomen, heliconians	600	worldwide
Pseudergolinae	constables, tabbies	20	India to Japan and Malay Archipelago
Apaturinae	emperors	100	worldwide but few in Africa, none in New Zealand
Biblidinae	88s, crackers, castors, dandies	600	mainly the Americas, with some in Africa, Indian region and western Malay Archipelago
Cyrestinae	daggerwings, mapwings	45	Americas, Africa, India to Japan, Malay Archipelago to Solomon Islands
Nymphalinae	admirals, buckeyes, tortoiseshells, European map, crescents	400	Worldwide

Glossary

alkaloids class of nitrogenous plant chemicals, some poisonous (eg pyrrolizidines)

amino acids class of nitrogenous compounds that form building blocks of proteins

anal claspers see *prolegs*

androconia special scales of various sorts possessed by males of many Lepidoptera species; almost invariably involved with pheromone communication

antennae pair of sensitive mobile appendages arising from head

biogeography the science of the distribution of living beings, concerned with why organisms occur where they do and where they do not

cardiac glycosides plant compounds that can arrest heart muscle

chromosomes paired, thread-like structures in each cell nucleus that carry most of an individual's genetic information; one member of each pair is maternal in origin, the other paternal

chrysalis (or pupa) immobile stage during which metamorphosis takes place

compound eye visual organ, an eye composed of numerous separate facets, each with its own lens, typically found in most adult insects including all butterflies

crypsis state of being camouflaged in full view, not hidden from view

diapause period or state of suspended development

ecdysis (or moulting) casting exoskeleton to permit growth of next stage

ecdysone (or moulting hormone) insect hormone that induces ecdysis

exoskeleton external covering of insect that provides support and protection

extinction the complete loss of all members of a local population of a species, or the global loss of all individuals of a given evolutionary lineage (species, genus or higher group)

family a group of genera considered to be more closely related by descent to each other than to any other such group of genera (e.g. Papilionidae, Nymphalidae); families are often divided into smaller groups (e.g. subfamilies and tribes), in all cases with standard endings to denote rank (e.g. subfamily Papilioninae, tribe Nymphalini within the subfamily Nymphalinae)

final instar last larval (caterpillar) stage before forming pupa

first instar first larval (caterpillar) stage that hatches from egg

frass caterpillar dung

genome entire genetic material of an individual organism or species

genus a group of species considered to be more closely related by descent to each other than to any other group of species (e.g.

Papilio, Heliconius); sometimes, when very distinct, a genus includes only one known species (as with *Baronia brevicornis*, the archaic swallowtail)

glucosides major class of organic chemical compounds derived from glucose

gynandromorphs individuals that are part male, part female

haemolymph insect blood

hair-pencils brushes or tufts of androconia that can be fanned out

halved gynandromorphs individuals that are all male on one side, all female on the other

hill-topping a mate-locating strategy: males defend hill-top territories and wait for females

hybrid individual whose parents belong to different species or subspecies; species hybrids are usually infertile

hyper-parasitoids parasitoids that do not feed on the primary insect host but on the larvae of other parasitoids

imaginal buds cells set aside in early development that resume growth in pupa to make major parts of adult (imago), including wings and genitalia

instar larval (caterpillar) developmental stage, between successive moults

juvenile hormone inhibits development of adult characters during insect development

larva caterpillar

legs three pairs of jointed, articulated legs that arise from the three segments of the insect thorax (see also *prolegs*)

mandibles insect 'jaws' used to cut up food – in butterflies only functional in the caterpillar stage

metamorphosis transformation from larva (caterpillar) to adult completed during pupal stage

meconium semi-liquid waste, often brightly coloured, voided from anus of adult after emergence from pupa

micropyle hole(s) in egg shell through which sperm can pass

mimicry similarity between two individuals or species that confers an advantage on one or both through the two being perceived as functionally the same by a third organism; Batesian mimicry is the term for a relatively unprotected species gaining advantage by looking like defended species avoided by predators; Müllerian mimicry refers to similarity among well-protected species avoided by predators

moulting see *ecdysis*

mud-puddling habit of butterflies congregating, often in large numbers, on mud or wet sand

order a major group of organisms, typically including many families, all considered to be more closely related by descent to each other than to any other such group (e.g. Lepidoptera, the butterflies and moths)

parasitic wasps insects distantly related to social wasps that lay their eggs in or on other insects, where they hatch and then eat their hosts alive, often from within

parasitoids organisms with the same basic lifestyle as parasitic wasps, thus including parasitic flies of the family Tachinidae

pheromones chemicals released by an individual that can influence behaviour of other individuals of same species

pigments naturally occurring chemical compounds that confer colour, including melanins, flavonoids, pterins and papilochromes

pockets and **pouches** invaginations of hindwing that contain androconia, found in males of certain milkweed butterflies

proboscis coiled feeding tube of adult used to ingest fluids

prolegs fleshy legs that arise from the abdominal segments of caterpillars; the prolegs together with the anal claspers of the last abdominal segment are totally different in structure from the three pairs of 'true', jointed legs found on the thorax of both larvae and adults

pupa see *chrysalis*

pupal diapause suspension of development during pupal stage

pupal rape copulation effected before female has emerged from pupa, denying her mate choice

scales hollow, flattened, typically coloured sac-like outgrowths that cover the body and wings; scales are modified insect hairs

sexual dimorphism differences between male and female of same species, especially in size, form and colour, other than primary sex organs

simple eye light sensitive insect organ with a fixed lens not capable of forming a focused image; typically six such 'eyes' occur on either side of the head of a caterpillar

species a particular sort or kind of organism in which, if sexually reproducing, all individuals are typically inter-fertile (not always the case when species e.g. occupy huge areas of the Earth); species names have two parts – the genus name first, with a capital letter, with a second name to denote the particular species within the genus (e.g. *Pieris brassicae*); the two names together being the 'scientific name' of the species

spermatophore packet or sac in which sperm and various accessory substances are transferred to a female by copulation with a male

sphragis female genital plug made by males of certain species during copulation; intended to prevent other males impregnating the female

spinneret organ just behind and between jaws of caterpillar that produces silk

spumaline polysaccharide-rich material smeared over eggs by certain species

structural colours colours not produced by pigments, but by the structure of the exoskeleton, most often involving multiple layers within wing scales

subspecies geographical subdivision of a species: a race

taxonomist scientist concerned with the recognition, comparison, classification, naming and means of identification of species and groups of organisms

variation differences in size, colour, form or physiology that invariably occur between individuals, of genetic, developmental and/or environmental origin

vestibulum genital opening or vulva of female

Index

Further information

Further reading

Butterflies, Edmund B. Ford. HarperCollins, London, 2009 (first published 1945).

Butterflies of the World (new edn.), Rod & Ken Preston-Mafham. Facts on File Publications, New York, 2004.

Butterfly. A Photographic Portrait, Thomas Marent. Dorling Kindersley, London, 2008.

Do Butterflies Bite? Fascinating Answers to Questions about Butterflies and Moths, Hazel Davies & Carol A. Butler. Rutgers University Press, New Jersey, 2008.

Ecology of Butterflies in Europe, Josef Settele, Tim Schreeve, Martin Konvička & Hans Van Dyck (eds.). Cambridge University Press, Cambridge, 2009.

Ecology and Evolution Taking Flight: Butterflies as Model Systems, Carol L. Boggs, Ward B. Watt & Paul R. Ehrlich (eds). University of Chicago Press, Chicago, 2003.

Lepidoptera, Moths and Butterflies (Vols. 1 and 2), Niels P. Kristensen (ed.). Series: *Handbook of Zoology*, Walter de Gruyter, Berlin, 1999, 2003.

The Biology of Butterflies (paperback edition), Richard I. Vane-Wright & Phillip R. Ackery (eds.). Princeton University Press, New Jersey, 1989.

The Butterflies of North America (paperback edn.), James A. Scott. Stanford University Press, California, 1992.

The Development and Evolution of Butterfly Wing Patterns, H. Frederik Nijhout. Smithsonian Institution Press, Washington, 1991.

The Lepidoptera (reprinted, with corrections), Malcolm J. Scoble. Oxford University Press, Oxford, 1995.

The Millennium Atlas of Butterflies in Britain and Ireland, Jim Asher, Martin Warren, Richard Fox, Paul Harding, Gail Jeffcoate & Stephen Jeffcoate. Oxford University Press, Oxford, 2001.

The Monarch Butterfly, Karen S. Oberhauser & Michelle J. Solensky (eds.). Cornell University Press, Ithaca, 2004.

Internet sources

There are numerous websites devoted to butterflies, many of which are primarily focused on nomenclature, illustration and identification. There are far fewer websites offering good and extensive information about butterfly biology. Please note that web addresses are subject to change.

British and Irish Butterflies and Moths: the Cockayne collection. Phase 1, Butterflies. http://www.nhm.ac.uk/research-curation/scientific-resources/biodiversity/uk-biodiversity/cockayne/

Butterflies and Moths of North America. http://www.butterfliesandmoths.org/

Hesperiidae Latreille 1809 Skippers. Tree of Life Project. http://tolweb.org/Hesperiidae/12028

HOSTS – a database of the world's Lepidoptera hostplants.
http://www.nhm.ac.uk/research-curation/research/projects/hostplants/

Learn About Butterflies - a guide to the world of butterflies and moths
http://www.learnaboutbutterflies.com/

Papilionoidea Latreille 1802 True Butterflies. Tree of Life Project. http://tolweb.org/Papilionoidea/12027

Conservation societies

Butterfly Conservation, Manor Yard, East Lulworth, Wareham, Dorset BH20 5QP, UK.
http://www.butterfly-conservation.org
[The aim of Society is to try to halt the decline of butterflies and at the same time help safeguard the environment itself. Specifically it aims to conserve both butterflies and moths, as well as the habitats on which they depend.]

Butterfly Conservation Europe, Postbus 506, NL-6700 AM Wageningen, The Netherlands.
http://www.bc-europe.eu/contact.php
[The association aims to prevent the extinction of any species of butterfly and moth, especially in Europe, and promote all relevant activities and initiatives.]

The Xerces Society for Invertebrate Conservation, 4828 SE Hawthorne Blvd., Portland, Oregon 97215, USA.
http://www.xerces.org
[An international non-profit organization dedicated to protecting biological diversity through conservation of invertebrates, including butterflies.]

Lepidoptera societies

Societas Europaea Lepidopterologica
http://www.soceurlep.eu/index.php?id=4

The Lepidopterists' Society of Africa
http://www.lepsoc.org.za

North American Butterfly Association
http://www.naba.org/

The Butterfly Society of Japan (Teinopalpus)
http://www.asahi-net.or.jp/~EY4Y-TKNM/bsjn/bsjn_e.html

The Lepidopterists' Society
http://www.lepsoc.org/

For many local societies see
http://butterflywebsite.com/society/list.htm

Acknowledgements

The author acknowledges almost countless 'aurelians' and other entomologists, both living and dead, without whose work this book would not have been possible. For inspiration I thank Michael Boppré, Lincoln Brower, Paul Ehrlich and Larry Gilbert, together with the late Dietrich Schneider, Miriam Rothschild, Suguru Igarashi, Ebbe Nielsen and Bob Silberglied. For practical help I am also very grateful to many friends and colleagues, including Hirotaka Matsuka, Yasutaka Murata, Osamu Yata, Phil DeVries, Gerardo Lamas, Larry Gilbert, Bob Robbins, Michael Boppré , Konrad Fiedler, Rienk de Jong, Mike Bascombe, Jim Mallet, Jeremy Thomas, Michael Usher, Phillip Ackery, George Beccaloni, David Carter, Bernard d'Abrera, Kim Goodger, Martin Honey, Shayleen James, David Lees, Geoff Martin, Klaus Sattler, Malcolm Scoble, Campbell Smith, John Tennent, Shen-Horn Yen, Frank Greenaway and Harry Taylor.

Picture credits